THE SUN
SHINES BRIGHT

Science Essay Collections by Isaac Asimov
From Fantasy and Science Fiction

From Other Sources

THE SUN
SHINES BRIGHT

Isaac Asimov

DOUBLEDAY & COMPANY, INC.
Garden City, New York
1981

The following essays in this volume are reprinted from *The Magazine of Fantasy and Science Fiction*, having appeared in the indicated issues:

"Out, Damned Spot!" March 1979
"A Long Day's Journey," April 1979
"The Inconstant Moon," May 1979
"Alas, All Human," June 1979
"Below the Horizon," July 1979
"Clone, Clone of My Own," August 1979
"How Little?" September 1979
"Just Thirty Years," October 1979
"The Sun Shines Bright," November 1979
"The Useless Metal," December 1979
"Neutrality!" January 1980
"The Finger of God," February 1980
"The Noblest Metal of Them All," March 1980
"Nice Guys Finish First!" April 1980
"More Crowded!" May 1980
"The Unsecret Weapon," June 1980
"Siriusly Speaking," July 1980

Dedicated to
Carol Bruckner and all the other nice people
at the Harry Walker lecture agency

CONTENTS

THE ELEMENTS

THE CELL

THE SCIENTISTS

THE PEOPLE

INTRODUCTION

What do I do about titles? It's a problem that, perhaps, I shouldn't plague you with, but I like to think that my Gentle Readers are all my friends, and what are friends for if not to plague with problems?

Many's the time I've sat staring at a blank sheet of paper for many minutes, unable to start a science essay even though I knew exactly what I was going to discuss and how I was going to discuss it and everything else about it—except the title. Without a title, I can't begin.

It gets worse with time, too, for I suffer under the curse of prolificity. Over two hundred thirty books; over three hundred short stories; over thirteen hundred nonfiction essays—and every one of them needing a title—a new title—a meaningful title—

Sometimes I wish I could just number each product the way composers do. In fact, I did this on two occasions. My hundredth and my two hundredth books are called *Opus 100* and *Opus 200* respectively. Guess what I intend to call my three hundredth book, if I survive to write it?

Numbers won't work in general, however. They look unlovely as titles (*1984* is the only successful example I can think of). They're hard to differentiate and identify. Imagine going into a bookstore and at the last minute failing to remember whether it is 123 or 132 you're looking for. I've met people who had trouble remembering the title of a book on calculus that was entitled *Calculus*.

Besides, editors insist on significant titles, and the sales staff insists on titles that sell, and I insist on titles that amuse me. Pleasing everybody is difficult, so I concentrate first on pleasing *me*.

There are several types of titles that please me where my individual science essays are concerned. I like quotations, for instance, which apply to the subject matter of the essay in an unexpected way.

For instance, we know exactly what Lady Macbeth meant when she cried out in agony, during her sleep-walking scene, "Out, damned spot!" but you could also say it to a dog named spot that had just walked onto the living room carpet with muddy feet, or you could apply it perfectly accurately as I did in my first essay.

And when Juliet warns Romeo against swearing by "the inconstant moon," she doesn't quite mean what I mean in the title of the ninth essay.

Another way of using a quotation is to give it a little twist. Leo Durocher said, "Nice guys finish last" and Mark Antony referred to Brutus as "the noblest Roman of them all." If I change a word to make a title that fits the subject matter of the essay, I am happy. Or I can change a cliché into its opposite and go from a "secret weapon" to an "unsecret weapon."

But I can't always. Sometimes I have to use something as pedestrian as "Neutrality!" or "More Crowded!" and then I am likely to write the entire essay with my lower lip trembling and my blue eyes brimming with unshed tears.

Even my science-essay collections have become numerous enough to cause me problems. This one is the fifteenth in a series taken from *The Magazine of Fantasy and Science Fic-*

tion (not counting four books which are reshufflings of essays in older volumes).

The first book in the series was entitled *Fact and Fancy* because, logically enough, the essays dealt with scientific fact (as understood at the time of writing) and with my own speculations on those facts.

The second and third books were entitled *View from a Height* and *Adding a Dimension* respectively. In each case, the title was a phrase taken from the introduction.

The third title gave me an idea, however. Why not, in each title, use a different word that is associated with science. The third title included the word "dimension," for instance.

The fourth title, therefore, became *Of Time and Space and Other Things*, which had the words "time" and "space" in it and which was (more or less) a description of the nature of the essays. After that, the titles included successively "earth," "science," "solar system," "stars," "electron," "moon," "matter(s)," "planet," "quasar" and "infinity."

Doubleday & Company, my esteemed publishers, did not altogether trust my colorful titles. They subtitled the first in the series "Seventeen Speculative Essays" on the book jacket, though not on the title page. They continued ringing changes on "essays on science" in the first five books in the series and then gave up and let the names stand by themselves. Sales were not adversely affected when the subtitles were omitted.

The title of the eighth book was *The Stars in Their Courses* which happened to be the title of one of the essays in the book.

That struck my fancy. Not every essay title is suitable for the entire collection, but out of seventeen essays at least one is very likely to be useful. It came about, then, that the eighth to fourteenth volumes inclusive (except for *Of Matters Great and Small*) each had titles duplicating that of one of the essays.

That brings us to this volume.

Some of the individual essay titles in this volume are obviously unsuitable for the book as a whole. To call the book *How Little?* or *Just Thirty Years* would give no idea at all as to the contents and that is unsporting.

To call it *The Finger of God* or *Nice Guys Finish First* would give an actively wrong view of the contents. I wouldn't want people to think the book dealt with either theology or self-improvement.

The Inconstant Moon would be a good title, but one of my essay volumes is already called *The Tragedy of the Moon.*

I was strongly tempted by *Clone, Clone of My Own,* but clones are a subject of such interest to the general public right now that many people who have never heard of me might be tempted to buy the book on the basis of the title and they would then be disappointed.

So that brought it down to *The Sun Shines Bright.* There is a slight flaw there in that the word "bright" occurs also in *Quasar, Quasar, Burning Bright,* but I have not used the word "sun" in any of the titles and it deserves a play, so I decided on that as the title.

Just remember, though, that the book has nothing to do with Kentucky, or with Stephen Foster.

THE SUN
SHINES BRIGHT

THE SUN

1.
Out, Damned Spot!

I love coincidences! The more outrageous they are, the better. I love them if only because irrationalists are willing to pin so many garbage-filled theories on them, whereas I see them only for what they are—coincidences.

For instance, to take a personal example . . .

Back in 1925, my mother misrepresented my age for a noble motive. She told the school authorities I had been born on September 7, 1919, so that on September 7, 1925, I would be six years old and would qualify to enter the first grade the next day (for which I was more than ready).

Actually, I was born on January 2, 1920, and was not eligible for another half year, but I was born in Russia and there were no American birth certificates against which to check my mother's statement.

In the third grade, I discovered that the school records had me down for a September 7 birthday and I objected so strenuously that they made the change to the correct January 2, 1920.

Years later, during World War II, I worked as a chemist at the U. S. Navy Yard in Philadelphia (along with Robert Heinlein and L. Sprague de Camp, as it happens), and that meant I was draft-deferred.

As the war wound down, however, and my work grew less important in consequence, the gentlemen of my draft board looked at me with an ever-growing yearning. Finally, five days after V-J day, I received my induction notice and · eventually attained the ethereal status of buck private.

That induction notice came on September 7, 1945, and at that time, only men under twenty-six years of age were being drafted. Had I not corrected my mother's misstatement of twenty years before, September 7 would have been my 26th birthday and I would not have been drafted.

But that is just a tiny coincidence. I have just come across an enormous one involving a historical figure—an even less likely one, I think, than I have recorded in connection with Pompey.[1] I will, of course, start at the beginning.

In medieval times, the scholars of Western Europe went along with Aristotle's dictum that the heavenly bodies were unchanging and perfect. In fact, it must have seemed that to believe anything else would have been blasphemous since it would seem to impugn the quality of God's handiwork.

In particular, the sun seemed perfect. It was a container suffused with heavenly light and it had not changed from the moment of its creation. Nor would it change at any time

---

[1] See "Pompey and Circumstance," in *Left Hand of the Electron* (Doubleday, 1972).

in the future until the moment it pleased God to bring the sun to an end.

To be sure, every once in a while the sun could be looked at with impunity when it shone through haze near the horizon and then it appeared, at rare moments, as though there were some sort of spot on it. This could be interpreted as a small dark cloud, or, perhaps, the planet Mercury passing between the sun and earth. It was never thought to be an actual flaw in the sun, which was, by definition, flawless.

But then, toward the end of 1610, Galileo used his telescope to observe the sun during the sunset haze (a risky procedure which probably contributed to Galileo's eventual blindness) and saw dark spots on the sun's disk every time. Other astronomers, quickly learning to make use of telescopes, also reported these spots and one of them was a German astronomer, Christoph Scheiner, who was a Jesuit.

Scheiner's superior, on hearing of the observation, warned Scheiner against trusting his observations too far. Aristotle had, after all, made no mention of such spots and that meant they could not exist.

Scheiner therefore published his observations anonymously and said they were small bodies that orbited the sun and were not part of it. In that way, he held to the Aristotelian dictum of solar perfection.

Galileo, who was short-tempered and particularly keen on retaining credit, argued the matter intemperately and, as was his wont, with brilliant sarcasm. (This aroused Jesuit hostility, which did its bit in bringing on Galileo's troubles with the Inquisition.)

Galileo insisted on his own observations being earlier and ridiculed the suggestion that the spots were not part of the sun. He pointed out that at either limb of the sun, the spots moved more slowly and were foreshortened. He therefore deduced that the spots were part of the solar surface, and that their motion was the result of the sun's rotation on its axis in a period of twenty-seven days. He was quite correct in this, and the notion of solar perfection died, to the chagrin of many in power, and this contributed to Galileo's eventual troubles, too.

After that, various astronomers would occasionally report sunspots, or lack of sunspots, and draw sketches of their appearance and so on.

The next event of real interest came in 1774, when a Scottish astronomer, Alexander Wilson, noted that a large sunspot, when approaching the limb of the sun so that it was seen sideways, looked as though it were concave. He wondered whether the dim borders of the sunspot might not be declivities, like the inner surface of a crater, and whether the dark center might not be an actual hole into the deeper reaches of the sun.

This view was taken up, in 1795, by William Herschel, the foremost astronomer of his time. He suggested that the sun was an opaque cold body with a flaming layer of gases all about it. The sunspots, by this view, were holes through which the cold body below could be seen. Herschel speculated that the cold body might be inhabited.

That turned out to be all wrong, of course, since, as it happens, the shining surface of the sun is its coldest part. The farther one burrows into the sun, the hotter it gets, until at the center the temperature is some fifteen million degrees. That, however, was not understood until the nineteen-twenties. Even the thin gases high above the solar surface are hotter than the shining part we see, with temperatures in excess of a million degrees, though that was not understood until the nineteen-forties.

As for sunspots, they are not really black. They are a couple of thousand degrees cooler than the unspotted portion of the sun's surface so that they radiate less light and look black by comparison. If, however, Mercury or Venus moves between us and the sun, each shows up on the solar disk as a small, *really* black circle, and if that circle moves near a sunspot, it can then be seen that the spot is not truly black.

Still, though the Wilson-Herschel idea was wrong, it roused further interest in sunspots.

The real breakthrough came with a German named Heinrich Samuel Schwabe. He was a pharmacist and his hobby was astronomy. He worked all day, however, so he could not very well sit up all night long looking at the stars. It occurred to him that if he could think up some sort of daytime astronomical task, he could observe during the slow periods at the shop.

A task suggested itself. Herschel had discovered the planet Uranus, and every astronomer now dreamed of discovering a planet. Suppose, then, there were a planet closer to the sun than Mercury was. It would always be so near the sun that it would be extremely difficult to detect it. Every once in a while, though, it might pass between the sun and ourselves. Why not, then, watch the face of the sun for any dark, moving circle?

It would be a piece of cake, if the spot were seen. It couldn't be a sunspot, which would not be perfectly round and would not travel across the face of the sun as quickly as a planet would. Nor could it be Mercury or Venus, if those two planets were known to be located elsewhere. And anything but Mercury, Venus or a sunspot, would be a new planet.

In 1825, Schwabe started observing the sun. He didn't find any planet, but he couldn't help noting the sunspots. After a while he forgot about the planet and began sketching the sunspots, which changed in position and shape from day to day. He watched old ones die and new ones form and he spent no less than seventeen years(!) observing the sun on every day that wasn't completely cloudy.

By 1843, he was able to announce that the sunspots did not appear utterly at random. There was a cycle. Year after

year there were more and more sunspots till a peak was
reached. Then the number declined until they were almost
gone and a new cycle started. The length of time from peak
to peak was about ten years.

Schwabe's announcement was ignored until the better-
known scientist Alexander von Humboldt referred to it, in
1851, in his book *Kosmos,* a large overview of science.

At this time, the Scottish-German astronomer Johann von
Lamont was measuring the intensity of earth's magnetic
field and had found that it was rising and falling in regular
fashion. In 1852, a British physicist, Edward Sabine,
pointed out that the intensity of earth's magnetic field was
rising and falling in time with the sunspot cycle.

That made it seem that sunspots affected the earth, and
so they began to be studied with devouring interest.

Each year came to be given a "Zürich sunspot number"
according to a formula first worked out in 1849 by a Swiss
astronomer, Rudolf Wolf, who was, of course, from Zürich.
(He was the first to point out that the incidence of auroras
also rose and fell in time to the sunspot cycle.)

Reports antedating Schwabe's discovery were carefully
studied and those years were given sunspot numbers as
well. We now have a sawtooth curve relating the sunspot
number to the years for a period of two and a half centuries.
The average interval between peak and peak over that time
is 10.4 years. This does not represent a metronomelike regu-
larity by any means, though, since some peak-to-peak intervals are as short as 7 years and some are as long as 17 years.

What's more, the peaks are not all equally high. There
was a peak in 1816 with a sunspot number of only about 50.
On the other hand, the peak in 1959 had a sunspot number
of 200. In fact, the 1959 peak was the highest recorded. The
next peak, in 1970, was only half as high.

Sunspots seem to be caused by changes in the sun's magnetic field. If the sun rotated in a single piece (as the earth

or any solid body does), the magnetic field might be smooth and regular and be contained largely below the surface.

Actually, the sun does not rotate as a single piece. Portions of the surface farther from its equator take longer to make a complete turn than do portions near the equator. This results in a shear-effect which seems to twist the magnetic lines of force, squeezing them upward and out the surface.

The sunspot appears at the point of emergence of the magnetic lines of force. (It was not till 1908, three centuries after the discovery of sunspots, that the American astronomer George Ellery Hale detected a strong magnetic field associated with sunspots.)

Astronomers have to work out reasons why the magnetic field waxes and wanes as it does; why the period varies in both length and intensity; why the sunspots first appear at a high latitude at the beginning of a cycle and work their way closer to the sun's equator as the cycle progresses; why the direction of the magnetic field reverses with each new cycle and so on.

It isn't easy, for there are a great many factors involved, most of which are ill understood (rather like trying to predict weather on the earth), but there's no reason why, in the end, it shouldn't be worked out.

Of course, the changing magnetic field of the sun produces changes in addition to the varying presences and positions of sunspots. It alters the incidence of the solar flares, the shape of the corona, the intensity of the solar wind and so on. None of these things have any obvious interconnection, but the fact that all wax and wane in unison makes it clear that they must have a common cause.

Changes in the intensity of the solar wind affect the incidence of auroras on earth, and of electrical storms, and probably alter the number and nature of the ionic seeds in the atmosphere about which raindrops can form. In that way, the weather can be affected by the sunspot cycle, and,

in consequence, the incidence of drought, of famine, of political unrest, might all be related to the sunspot cycle by
enthusiasts.

In 1893, the British astronomer, Edward Walter Maunder,
checking through early reports in order to set up data for
the sunspot cycle prior to the eighteenth century, was astonished to find that there were virtually no reports on
sunspots between the years 1643 and 1715. (These boundary years are arbitrary to some extent. The ones I have
chosen—for a hidden reason of my own, which I will reveal
later—are just about right, however.)

There were fragmentary reports on numerous sunspots
and even sketches of their shapes in the time of Galileo and
of his contemporaries and immediate successors, but after
that there was nothing. It wasn't that nobody looked. There
were astronomers who did look and who reported that they
could find no sunspots.

Maunder published his findings in 1894, and again in
1922, but no one paid any attention to him. The sunspot
cycle was well established and it didn't seem possible that
anything would happen to affect it. An unspotted sun was
as unacceptable in 1900 as a spotted sun had been in 1600.

But then, in the nineteen-seventies, the astronomer John
A. Eddy, coming across the report of what he eventually
called the "Maunder minimum," decided to look into the
matter.

He found, on checking, that Maunder's reports were correct. The Italian-French astronomer Giovanni Domenico
Cassini, who was the leading observer of his day, observed a
sunspot in 1671 and wrote that it had been twenty years
since sunspots of any size had been seen. He was astronomer enough to have determined the parallax of Mars and to
have detected the "Cassini division" in Saturn's rings, so he
was surely competent to see sunspots if there were any. Nor
was he likely to be easily fooled by tales that there weren't
any if those tales were false.

John Flamsteed, the Astronomer Royal of England, another very competent and careful observer, reported at one time that he had finally seen a sunspot after seven years of looking.

Eddy investigated reports of naked-eye sighting of sunspots from many regions, including the Far East—data which had been unavailable to Maunder. Such records go back to the fifth century B.C. and generally yield five to ten sightings per century. (Only very large spots can be seen by the naked eye.) There are gaps, however, and one of those gaps spans the Maunder minimum.

Apparently, the Maunder minimum was well known till after Schwabe had worked out the sunspot cycle and it was then forgotten because it didn't fit the new knowledge. As a matter of fact, it may have been because of the Maunder minimum that it took so long after the discovery of sunspots to establish the sunspot cycle.

Nor is it only the reports of lack of sunspots that establish the existence of the Maunder minimum. There are reports consistent with it that deal with other consequences of the sun's magnetic field.

For instance, it is the solar wind that sets up auroras, and the solar wind is related to the magnetic field of the sun, particularly to the outbursts of energetic solar flares, which are most common when the sun is most magnetically active —that is, at times of high sunspot incidence.

If there were few if any sunspots over a seventy-year period, it must have been a quiet time generally for the sun, from a magnetic standpoint, and the solar wind must have been nothing but a zephyr. There should have been few if any auroras visible in Europe at that time.

Eddy checked the records and found that reports of auroras were indeed just about absent during the Maunder minimum. There were many reports after 1715 and quite a few before 1640, but just about none in between.

Again, when the sun is magnetically active, the lines of

force belly out from the sun with much greater strength than they do when it is magnetically inactive. The charged particles in the sun's outer atmosphere, or corona, tend to spiral about the lines of force, and do so in greater numbers, and more tightly, the stronger the lines of force are.

This means that the appearance of the corona during a total eclipse of the sun changes according to the position of the sun in the sunspot cycle. When the number of sunspots is near its peak and the magnetic activity of the sun is high, the corona is full of streamers radiating out from the sun and it is then extraordinarily complex and beautiful.

When the number of sunspots is low, there are few if any streamers and the corona seems like a rather featureless haze about the sun. It is then not at all remarkable.

Unfortunately, during the Maunder minimum, it was not yet the custom for astronomers to travel all over the world to see total eclipses (it wasn't as easy then, as it became later, to travel long distances), so that only a few of the over sixty total eclipses of the period were observed in detail. Still, those that were observed showed coronas that were, in every case, of the type associated with sunspot minima.

The auroras and the corona are bits of entirely independent corroboration. There was no reason at the time to associate them one way or another with sunspots, and yet all three items coincide as they should.

One more item, and the most telling of all:

There is always some radioactive carbon-14 in atmospheric carbon dioxide. It is produced by cosmic rays smashing into nitrogen atoms in the atmosphere. Plants absorb carbon dioxide and incorporate it into their tissues. If there happens to be more carbon-14 than usual in the atmospheric carbon dioxide in a particular year, then, in that year, the plant tissue that is laid down is richer than normal in that radioactive atom. The presence of carbon-14, whether slightly more or slightly less than normal, is always exceedingly tiny, but radioactive atoms can be detected with great delicacy and precision and even traces are enough.

Now it happens that when the sun is magnetically active, its magnetic field bellies so far outward that the earth itself is enveloped by it. The field serves to deflect some of the cosmic rays so that less carbon-14 is formed and deposited in plant tissues.

When the sun's magnetic field shrinks at the time of sunspot minima, the earth is not protected, so that more cosmic rays strike and more carbon-14 is formed and deposited.

In short, plant tissues formed in years of sunspot minima are unusually high in carbon-14, while plant tissues formed in years of sunspot maxima are unusually low in carbon-14.

Trees lay down thicknesses of wood from year to year, and these are visible as tree rings. If we know the year when a tree was cut down and count the rings backward from the bark, one can associate any ring with a particular year.

If each tree ring is shaved off and is separately analyzed for its carbon-14 content (making allowance for the fact that the carbon-14 content declines with the years as the atoms break down at a known rate), one can set up a sunspot cycle without ever looking at the solar records. (This is a little risky, of course, since there may be other factors that raise and lower the carbon-14 content of atmospheric carbon dioxide in addition to the behavior of the sun's magnetic field.)

As it happens, tree rings dating from the second half of the seventeenth century are indeed unusually high in carbon-14, which is one more independent confirmation of the Maunder minimum.

In fact, tree-ring data are better than anything else for two reasons. In the first place, they do not depend on the record of human observations, which is, naturally, subjective and incomplete. Secondly, whereas human observations are increasingly scanty as we move back in time before 1700, tree-ring data are solid for much longer periods.

In fact, if we make use of bristlecone pines, the living objects with the most extended lifetimes, we can trace back the variations in carbon-14 for five thousand years; in short, throughout historic times.

Eddy reports that there seem to be some twelve periods over the last five thousand years in which solar magnetic activity sank low; the extended minima lasting from fifty to a couple of hundred years. The Maunder minimum is only the latest of these.

Before the Maunder minimum there was an extended minimum from 1400 to 1510. On the other hand there were periods of particularly high activity such as one between 1100 and 1300.

Apparently, then, there is a long-range sunspot cycle on which the short-range cycle discovered by Schwabe is superimposed. There are periods when the sun is quiet and the magnetic field is weak and well behaved and the sunspots and other associated phenomena are virtually absent. Then there are periods when the sun is active and the magnetic field is undergoing wild oscillations in strength so that sunspots and associated phenomena reach decennial peaks.

What causes this long-range oscillation between Maunder minima and Schwabe peaks?

I said earlier that the sunspots seem to be caused by the differential rotation of different parts of the solar surface. What, then, if there were no difference in rotation?

From drawings of sunspots made by the German astronomer Johannes Hevelius in 1644, just at the beginning of the Maunder minimum, it seems that the sun may have been rotating all in one piece at that time. There would therefore be no shear, no twisted magnetic lines of force, nothing but a quiet, well-behaved magnetic field—a Maunder minimum.

But what causes the sun periodically to turn in one piece

and produce a Maunder minimum and then to develop a differential rotation and produce a Schwabe peak?

I'm glad to be able to answer that interesting question clearly and briefly: No one knows.

And what happens on earth when there is a Maunder minimum? As it happens, during that period Europe was suffering a "little ice age," when the weather was colder than it had been before or was to be afterward. The previous extended minimum from 1400 to 1510 also saw cold weather. The Norse colony in Greenland finally died out under the stress of cold after it had clung to existence for over four centuries.

But that may be only coincidence, and I have a better one.

What is the chance that a monarch will reign for seventy-two years? Obviously very little. Only one monarch in European history has managed to reign that long, and that was Louis XIV of France.

Given a reign of that length, and a Maunder minimum of that length, what are the odds against the two matching exactly? Enormous, I suppose, but as it happens, Louis XIV ascended the throne on the death of his father in 1643 and remained king till he died in 1715. He was king precisely through the Maunder minimum.

Now, in his childhood, Louis XIV had been forced to flee Paris to escape capture by unruly nobles during the civil war called the Fronde. He never forgave either Paris or the nobles.

After taking the reins of government into his own hands upon the death of his minister, Jules Mazarin, in 1661, Louis decided to make sure it would never happen again. He planned to leave Paris and build a new capital at Versailles in the suburbs. He planned to set up an elaborate code of etiquette and symbolism that would reduce the proud nobility into a set of lackeys who would never dream of rebelling.

He would, in short, make himself the unrivaled symbol of the state ("I am the state," he said), with everyone else shining only by the light of the king.

He took as his symbol, then, the unrivaled ruler of the solar system, the sun, from which all other bodies borrowed light. He called himself Le Roi Soleil.

And so it happened that the ruler whose long reign exactly coincided with the period when the sun shone in pure and unspotted majesty—something whose significance could not possibly have been understood at the time—called himself, and is still known as the Sun King.

2.

The Sun Shines Bright

As you all know, I like to start at the beginning. This occasionally upsets people, which is puzzling.

After all, the most common description I hear of my writing is that "Asimov makes complex ideas easy to understand." If that is so, might it not have something to do with the fact that I start at the beginning?

Yet editors who are publishing my material for the first time sometimes seem taken aback by a beginning at the beginning and ask for a "lead."

Even editors who have had experience with me sometimes feel a little uneasy. I was once asked to write a book about the neutrino, for instance, and I jumped at the chance. I even thought up a catchy title for it. I called it *The Neutrino*.

I began the book by describing the nature of the great generalizations we call the laws of nature. I talked about things like the conservation of energy, the conservation of momentum and so on. I pointed out that these laws were so useful that when an observed phenomenon went against one of them, it was necessary to make every reasonable effort to make the phenomenon fit the law before scrapping the whole thing and starting over.

All this took up precisely half the book. I was then ready to consider a certain phenomenon that broke not one conservation law but three of them, and pointed out that by postulating the existence of a particle called the neutrino, with certain specified properties, all three conservation laws could be saved at one stroke.

It was because I had carefully established the foundation that it would be possible to introduce the neutrino as an "of course" object with everyone nodding their heads and seeing nothing mysterious in supposing it to exist, or in the fact that it was only detected twenty-five years after its existence had been predicted.

With considerable satisfaction, I entitled Chapter 7 "Enter the Neutrino."

And, in the margin, my editor penciled, "At last!!!"

So now I will consider some aspects of the neutrino that have achieved prominence after I wrote that book. And again, I warn you it will take me a little time to get to the neutrino.

The sun shines bright because some of its mass is continually being converted into energy. In fact, the sun, in order to continue to shine in its present fashion, must lose 4,200,000,000 kilograms of mass every second.

At first blush, that makes it seem as though the sun doesn't have long for this universe. *Billions* of kilograms every *second?*

There are just about 31,557,000 seconds in one year and the sun has been shining, in round numbers, for

5,000,000,000 years. This means that in its lifetime (if we assume it has been shining in precisely the same way as it now is for all that time) the sun must have lost something like 158,000,000,000,000,000 kilograms of mass altogether.

In that case, why is it still here? Because there's so much of it, that's why.

All that mass loss I have just described, over its first 5 billion years of existence, represents only one ten-trillionth of the total mass of the sun. If the sun were to continue losing mass in this fashion and if it were to continue shining as it does today in consequence, it would last (if mass loss were the only requirement) for over 60 billion trillion years before snuffing out like a candle flame.

The trouble is, the sun isn't simply losing mass; it is doing so as the result of specific nuclear reactions. These nuclear reactions take place in a fairly complicated manner, but the net result is that hydrogen is converted to helium. To be more specific, four hydrogen nuclei, each one consisting of a single proton, are converted into a single helium nucleus consisting of two protons and two neutrons.

The mass of a proton is (in the standard units of mass used today) 1.00797, and four of them would consequently have a mass of 4.03188. The mass of a helium nucleus is 4.00260. In converting four hydrogen nuclei into a helium nucleus, there is thus a loss of 0.0293 units of mass, or 0.727 per cent of the mass of the four protons.

In other words, we can't expect the sun to lose all its mass when all the hydrogen is gone. It will lose only 0.727 per cent of its mass as all the hydrogen is converted into helium. (It can lose a bit more mass by converting helium into still more complicated nuclei, but this additional loss is small in comparison to the hydrogen-to-helium loss and we can ignore it. We can also ignore the small losses involved in maintaining the solar wind.)

Right now, in order for it to shine bright, the sun is converting 580,000,000,000 kilograms of hydrogen into helium every second.

If the sun had started its life as pure hydrogen and if it consumed hydrogen at this same steady rate always, then its total lifetime before the last dregs of hydrogen were consumed would still be something like 100 billion years.

To be sure, we suspect that the sun was formed as something other than pure hydrogen. The composition of the original cloud that formed it seems to have already been 20 per cent helium. Even so, there seems to be enough hydrogen in the sun to keep it going for 75 billion years at its present rate.

And yet it won't continue that long at its present rate; not nearly. The sun will continue to shine in more or less its present fashion for only about 7 billion years perhaps. Then, at its core, which will be growing larger and hotter all that time, helium will start to fuse and this will initiate a series of changes that will cause the sun to expand into a red giant and, eventually, to collapse.

Even when it begins to collapse, there will still be plenty of hydrogen left. In fact, a star large enough to form a supernova shines momentarily as bright as a whole galaxy of stars because so much of the hydrogen it *still* possesses goes off all at once.

Clearly, if we are going to understand the future of the sun, we must know more than its content of hydrogen and the present rate of hydrogen loss. We must know a great deal about the exact details of what is going on in its core right now so that we may know what will be going on in the future.

Let's tackle the matter from a different angle. If four protons are converted to a two-proton-two-neutron helium nucleus, then two of the original protons must be converted to neutrons.

Of the 580,000,000,000 kilograms of hydrogen being turned to helium every second, half, or 290,000,000,000 kilograms represents protons that are being turned to neutrons.

There are, as it happens, just about 600,000,000,000,000,-000,000,000,000 protons in every kilogram of hydrogen, a figure it is easier to represent as 6×10^{26}. That means that there are, roughly, 1.75×10^{38} protons in 290,000,000,-000 kilograms; or, if you want it in an actual string: 175,000,000,000,000,000,000,000,000,000,000,000,000.

In the core of the sun then, 1.75×10^{38} protons are being converted to 1.75×10^{38} neutrons every second. That is what makes it possible for you to get a nice suntan on the beach; or if you want to be lugubrious about it, that is what makes it possible for life to exist.

A proton doesn't change to a neutron just like that, however. The proton has a positive electric charge and the neutron is uncharged. By the law of conservation of electric charge, that positive charge can't disappear into nothingness. For that reason, when a proton is converted to a neutron, a positron is also formed. The positron is a light particle, with only 1/1811 the mass of a proton, but it carries exactly the positive electric charge of a proton.

But then, the positron cannot be formed all by itself, either. It is a particle of a kind that exists in two varieties, "leptons" and "antileptons." If a particle of one of those varieties is formed, then a particle of the other variety must also be formed. This is called the law of conservation of lepton number. This conservation law comes in two varieties, the conservation of electron-family number and the conservation of muon-family number.[1]

The positron is an example of an antilepton of the electron family. We have to form a lepton of the electron family to balance it. The neutron and the positron, in forming, have consumed all the mass and electric charge in the original proton, so the balancing lepton must have neither mass nor charge. It must, however, have certain quantities of energy, angular momentum and so on.

---

[1] There might conceivably be an infinite number of other such lepton families each with its conservation law, but we needn't worry about that here.

The lepton that is formed to balance the positron is the massless, chargeless neutrino.

At the core of the sun, then, there are formed, every second, 1.75×10^{38} positrons and 1.75×10^{38} neutrinos.

We can ignore the positrons. They remain inside the sun, bouncing off other particles, being absorbed, re-emitted, changed.

The neutrinos, however, are a different matter. Without mass and without charge, they are not affected by three of the four types of interaction that exist in the universe—the strong, the electromagnetic and the gravitational. They are affected only by the weak interaction.

The weak interaction decreases in intensity so rapidly with increasing distance that the neutrino must be nearly in contact with some other particle in order to be influenced by that weak interaction. As it happens, though, the neutrino behaves as though it has a diameter of 10^{-21} centimeters, which is a hundred millionth the width of a proton or neutron. It can therefore slip easily through matter without disturbing it. And even if it does happen to approach an atomic nucleus, a neutrino is massless and therefore moving at the speed of light. Unlike the rather slow-moving protons and neutrons, a neutrino doesn't stay in the neighborhood of another particle for longer than 10^{-23} seconds.

The consequence is that a neutrino virtually never interacts with any other particle but streaks through solid matter as though it were a vacuum. A beam of neutrinos can pass through a light-year of solid lead and emerge scarcely attenuated.

This means that the neutrinos formed at the center of the sun are not absorbed, re-emitted or changed in any significant manner. Indifferent to their surroundings, the neutrinos move out of the sun's core in all directions, at the speed of light. In three seconds after formation, the neutrinos formed at the sun's core reach the sun's surface and move out into space. The sun is therefore emitting $1.75 \times$

10^{38} neutrinos into space every second and, presumably, in every direction equally.

In a matter of eight minutes after formation, these solar neutrinos are 150 million kilometers from the sun, and that happens to be the distance at which the earth orbits the sun.

Not all the solar neutrinos reach the earth, however, because not all happen to have been moving in the direction of the earth. The solar neutrinos can be envisaged, eight minutes after formation, as moving through a huge hollow sphere with its center at the sun's center and its radius equal to 150 million kilometers. The surface area of such a sphere is about 2.8×10^{17} square kilometers.

If the solar neutrinos are moving in all directions equally, then through every square kilometer of that imaginary sphere there are passing 6.3×10^{20} neutrinos. There are 10 billion (10^{10}) square centimeters in every square kilometer, so 6.3×10^{10} (63 billion) neutrinos pass through every square centimeter of that imaginary sphere every second.

Part of the sphere is occupied by the earth. The earth has a radius of 6378 kilometers, so that its cross-sectional area is roughly 128,000,000 square kilometers or about 1/2,000,000 of the total imaginary sphere surrounding the sun.

A total of about 80,000,000,000,000,000,000,000,000,000,000 solar neutrinos are passing through the earth every second, day and night, year in, year out.

And how many do you get? Well, a human being is irregular in shape. To simplify matters, let us suppose a human being is a parallelepiped who is 170 centimeters tall, 35 centimeters wide and 25 centimeters thick. The smallest cross section would be 35×25, or 875 square centimeters, and the greatest cross section would be 35×170, or 5950 square centimeters. The actual cross section presented by a human being to the neutrino stream would depend on his or her orientation with respect to the sun.

Let's suppose that 3400 square centimeters represents a

reasonable average cross section presented to the neutrino stream. In that case, a little over 200,000,000,000,000 (200 trillion) solar neutrinos are passing through your body every second—without bothering you in any way.

To be sure, every once in a while, a neutrino *will* just happen to strike an atomic nucleus squarely enough to interact and induce a nuclear reaction that would be the reverse of one that would have produced a neutrino. The conversion of a proton to a neutron produces a neutrino, so the absorption of a neutrino converts a neutron to a proton. The emission of a neutrino is accompanied by the emission of a positron. The absorption of a neutrino is accompanied by the emission of an electron, which is the opposite of a positron.

In the human body there may be one neutrino absorbed every fifty years, but physicists can set up more efficient absorbing mechanisms.

If a neutrino strikes a nucleus of chlorine-37 (17 protons, 20 neutrons), then one of the neutrons will be converted to a proton and argon-37 (18 protons, 19 neutrons), along with an electron, will be formed.

To make this process detectable, you need a lot of chlorine-37 atoms in close proximity so that a measurable number of them will be hit. Chlorine-37 makes up one fourth of the atoms of the element chlorine. As a gas, chlorine is mostly empty space, and to liquefy it and bring its two-atom molecules into contact requires high pressure, low temperature or both. It is easier to use perchloroethylene, which is a liquid at ordinary temperature and pressure, and which is made up of molecules that each contain two carbon atoms and four chlorine atoms. The presence of the carbon atoms does not interfere and perchloroethylene is reasonably cheap.

Of course, you want a lot of perchloroethylene; 100,000 gallons of it, in fact. You also want it somewhere where *only* neutrinos will hit it, so you put it a mile deep in a gold mine in South Dakota. Nothing from outer space, not even the strongest cosmic-ray particles, will blast through the mile of rock to get at the perchloroethylene. Nothing except neutrinos. They will slide through the rock as though it weren't there and hit the perchloroethylene.

What about the traces of radioactivity in the rocks all around the perchloroethylene? Well, you surround the vat with water to absorb any stray radioactive radiations.

In 1968, Raymond Davis, Jr., did all this and began capturing neutrinos. Not many. Every couple of days he would capture *one* in all those gallons of perchloroethylene. He would let the captures accumulate, then use helium gas to flush out any argon atoms that had formed. The few argon-37 atoms could be counted with precision because they are radioactive.

There was a surprise, though. Neutrinos were captured—but not enough. Davis got only one sixth of the neutrinos he expected in his early observations. After he plugged every last loophole and worked at it for ten years, he was able to get the number up to one third of what was expected, but not more.

But then it is exciting to have something unexpectedly go wrong!

If the experiment had worked perfectly, scientists would only know that their calculations were correct. They would be gratified but would be no further ahead.

Knowing that something is wrong means that they must return to the old drawing board, go over what it was they thought they knew. If they could modify their theory to explain the anomalous observation, they might find that the new (and presumably better) theory could, perhaps quite unexpectedly, explain other mysteries as well.

Yes, but how explain the anomaly?

All sorts of things are being suggested. Perhaps the theory of neutrino formation is wrong. Perhaps neutrinos aren't stable. Perhaps there are factors in the core of the sun, mixing effects or nonmixing effects, that we aren't taking into account. Perhaps the sun has even stopped working for some reason and eventually the change will reach the surface and it will no longer shine bright and we will all die.

In science, however, we try to find the *least* adjustment of theory that will explain an anomaly, so before we kill the sun, let's think a little.

According to our theories, the hydrogen doesn't change directly into helium. If that were so, all the neutrinos formed would be of the same energy. What does happen is that the hydrogen turns to helium by way of a number of changes that take place at different speeds, some of the changes representing alternate pathways. Neutrinos are produced at different stages of the process and every nuclear change that produces a neutrino produces one with a characteristic energy.

The result is that of the many billions of neutrinos constantly passing through any object, a certain percentage have this much energy, a certain percentage have that much and so on. There's a whole spectrum of energy distribution to the neutrinos, and the exact nature of the spectrum mirrors the exact details of the route taken from hydrogen to helium. Any change in the route will produce a characteristic change in the spectrum.

Naturally, the more energetic a neutrino, the more likely it is to induce a nuclear change and the perchloroethylene detects only the most energetic neutrinos. It detects only those produced by one particular step in the conversion of hydrogen to helium. That one particular step is the conversion of boron-8 to beryllium-8.

The neutrinos formed by any other reaction taking place in the overall hydrogen-helium conversion do not contribute significantly to the absorptions in the perchloroethylene tank. The deficiency in solar neutrinos detected by Davis is therefore a deficiency in the boron–beryllium conversion and nothing more.

How can we be sure that our theory is correct about the details of what is going on in the sun's core? How can we be sure that Davis should have observed three times as many neutrinos as he did?

We can't, after all, check how much boron-8 is actually present in the sun and how rapidly and energetically it breaks down to beryllium-8. Our theory concerning that depends on determining reaction rates under laboratory conditions and then extrapolating them to conditions at the sun's core. By working with these extrapolated reaction rates, we can calculate a number of reactions that one way or another contribute to the formation of boron-8 and in this way determine its overall concentration. But what if we're not extrapolating properly?

After all, the nuclear reaction rates may depend quite strongly on the temperature and pressure within the sun, and how sure can we be that we're not a bit off on the temperature or pressure or both?

In order to be able to talk sensibly about the neutrinos detected by Davis—whether they're too many, too few or just right—we really need to know more about the conditions at the core of the sun, and the only way we can do that more accurately than by long-range and difficult calculations from observations at laboratory conditions is to study the entire neutrino spectrum.

If we could study the entire neutrino spectrum, we might be able to deduce from that the various individual steps in the hydrogen-helium conversion, and the concentrations and breakdown speeds of all the various nuclear intermediates.

If this relatively direct knowledge of the sun's core doesn't gibe with the extremely indirect knowledge based on extrapolation from laboratory experiments, then we will have to accept the former, re-examine the latter and develop, perhaps, new concepts and new rules for nuclear reactions.

In short, instead of learning about the sun's core from our own surroundings, as we have been trying to do hitherto, we may end up learning about our own surroundings from the sun's core.

To get the full spectrum, we will need detecting devices other than perchloroethylene. We will need a variety of "neutrino telescopes."

One possibility is that of making use of gallium-71 (31 protons, 40 neutrons), which makes up 40 per cent of the element gallium as it occurs in nature. Neutrino absorption would convert it to radioactive germanium-71 (32 protons, 39 neutrons).

You would need about 50 tons of gallium-71 if you wanted to trap one solar neutrino per day. That is only one twelfth of the mass of the 100,000 gallons of perchloroethylene, but the gallium is much more than twelve times as expensive. In fact that much gallium would cost about $25 million right now.

Gallium is liquid at temperatures well below the boiling point of water, so that germanium-71 can be flushed out without too much trouble. The advantage of gallium over perchloroethylene is that gallium will detect neutrinos of lower energy than perchloroethylene will.

In 1977, Ramaswamy S. Raghavan at Bell Laboratories suggested something even more exciting, perhaps. He suggested that indium-115 (49 protons, 66 neutrons) be used as a neutrino absorber. Indium-115 makes up 96 per cent of the natural metal and when it absorbs a neutrino, it is converted to tin-115, which is stable. The tin-115, however, is produced in an excited (that is, high-energy) state and it gives up that energy and returns to normal by emitting two gamma rays of characteristic energies a few millionths of a second after being formed. In addition, there is the inevitable electron that is hurled out of the indium-115 nucleus.

The formation of an electron and two gamma rays at virtually the same time is, in itself, sufficient indication of neutrino capture and there would be no necessity to isolate the atoms of tin-115.

What's more, by measuring the energy of the electron hurled out of the indium-115 nucleus, one could determine the energy of the incoming neutrino. The indium detector could thus give us our first picture of the neutrino spectrum as a whole.

And more, too. After all, how do we really know the neutrinos detected by Davis came from the sun? Suppose there is some other source we're unaware of, and suppose we're getting nothing from the sun?

In the case of the indium detector, the fleeing electrons will move pretty much in line with the incoming neutrino. If the line of motion of the electron, extended backward, points toward the sun no matter what time of day it is, it will be a fair conclusion that the neutrinos are indeed coming from the sun.

Working up a system that will detect gamma rays and electrons and measuring the direction and energy of them won't be easy, but it probably can be done. About four tons of indium-115 would be needed to detect one neutrino a day and the overall cost might be $10 million.

It will take some years to set up these detection devices, but I feel that as neutrino telescopes are devised and improved, the resulting science of "neutrino astronomy" may end up revolutionizing our knowledge of the universe in the same way that light telescopes did after 1609 and radio telescopes did after 1950.

3.

The Noblest Metal of Them All

I was at lunch with a group of men yesterday in a pleasant midtown restaurant when, quite unexpectedly, a woman accosted me with great excitement and glee. She was white-haired, roughly my age and attractive.

What was very evident was that she was greeting me in the style of an old friend and, as is usual, a pang of exquisite embarrassment shot through me. I don't know why it is but though all my old friends seem to have no trouble remembering me, I have the devil's own time remembering them. A brain deficiency, I think, born of trying too hard never to

forget the names of all the elements and the distances of all
the planets.

I relaxed a trifle when it turned out from her ebullient
conversation that she was a friend of my sister's actually,
and that her only connection with us dated back to 1938.
Really, with a gap in time like that, difficulty in remembering is but a venial sin.

Then she said, "But I always knew, even then, Dr. Asimov, that you were going to be successful and famous someday."

The proper response, of course, would have been a modest simper and a shy hanging of the head, but another thing
I have the devil's own time remembering is the proper response.

Instead, I said, "If you knew that, then why didn't you
tell me?"

Actually, though, now that I think it over in cold blood, I
wouldn't have wanted her to tell me. The surprises that
time brings make up much of the excitement of life—and of
science.

Which, of course, brings me to the subject of this essay.

Gold is rare, it is beautiful, it is dense, it neither rusts nor
decays.

The rareness and beauty call for no comment, but we can
put figures to the density most dramatically by comparing it
with lead.

Lead is about three thousand times as common as gold in
earth's crust and is as ugly in its grayish coloring as gold's
gleaming yellow is beautiful. Lead is common enough for
day-to-day use, therefore, and valueless for anything else.

Lead is pretty dense, however, and since it is the densest
object ordinary people in ancient times were liable to come
across, it became a byword for density.

You walk with leaden feet when you are leaden-hearted,
or when your eyes are leaden-lidded for want of sleep.

Things lie heavy as lead on your bosom when you are unhappy.

Yet if the density of lead is 1, the density of gold is 1.7. If you have a lump of lead and a lump of gold of equal shape and size, and the lead weighs, let us say, 3 kilograms, the gold would weigh 5 kilograms. If being leaden-hearted is to be sorrowful and unhappy, imagine how sorrowful and unhappy you would be if you were golden-hearted—except that is not how metaphors work.

As soon as you use gold in your metaphors, it is the beauty and value that express themselves, not the density. Therefore, if you trudge heavily on leaden feet when you are miserable, you dance trippingly on golden feet when you are happy.

The permanence of gold rests on its very small tendency to combine with other kinds of atoms. It therefore does not rust, is not affected by water or other substances. It even remains untouched by most acids.

This resistance against the influence of other substances, this haughty exclusiveness, led people to speak of gold as a "noble metal," since it nobly scorns to associate with substances of lesser quality. The social metaphor was carried over to metals like lead and iron, which were not so incorruptible and were therefore "base" metals, where the "base" represents low position in social standing.

Now, then, what are the chances of there being metals that are nobler than gold, rarer, denser, less apt to change? To an ancient, the notion might have been a laughable one, since gold had so long been used metaphorically for perfection (even the streets of heaven could find no better paving blocks than gold). To ask for something nobler than gold would be to ask for something that improved on perfection.

And yet such a better-than-gold metal exists, is now well known, and was, in fact, sometimes found and used even in ancient times. It is found in a metal artifact in Egypt dating back to the seventh century B.C., and some of the Incan metal artifacts in pre-Columbian South America are made of an alloy of gold and this other metal.

The first specific reference to it in the scientific writings of Europeans came in 1557. An Italian scholar, Julius Caesar Scaliger (1484–1558), mentioned a metal found in Central America which could not be liquefied by any heat applied to it.

Here was immediately an indication that it surpassed gold in one respect. Of the metals known to the ancients, mercury melted at very low temperatures, and tin and lead at only moderately high ones. Of the other four, silver melted at 961° C, gold at 1063° C, copper at 1083° C, and iron at 1535° C.

One might have suspected that if gold were truly noble it would resist fire as well as air and water and would not melt. The fact that copper, which is baser than gold, melts at a slightly higher temperature, and that iron, which is considerably baser than gold, melts at a considerably higher temperature is rather disconcerting. (For all I know, it might have been viewed as a heavenly dispensation to permit iron to be hard and tough enough to be used for weapons of war, something too utilitarian for the nobility of gold.)

Clearly, the new metal must melt at temperatures higher than iron does.

The first scientists to study the metal and describe it in detail were an English metallurgist, Charles Wood, and a Spanish mathematician, Antonio de Ulloa (1716–95). Both, in the seventeen-forties, studied specimens that came from South America. One place where the new metal was obtained was as nuggets in the sands of the Pinto River in Colombia. Since the metal was whitish, the Spaniards on the spot called it "Pinto silver." They used the Spanish language, so that it was *platina del Pinto*.

Pinto silver was not real silver, of course. It was much denser than silver and it melted at a much higher temperature. It didn't even really look like silver. There is a distinct yellowish touch to silver which gives it a light, warm look that other white metals do not have. Aluminum and chromium may be white and shiny, but they are not silvery in appearance, and neither is *platina del Pinto*.

Eventually, when the "-um" ending became standard for metals, the "Pinto" portion of the name was dropped and the new metal became "platinum." In English, platinum and silver are so unlike in name that the connection is lost. In Spanish, however, silver is *plata* and platinum is *platina*.

Chemists became intensely interested in platinum after its discovery, but there wasn't much that could be done with it usefully. Either it had to be left in its original lump or it could be dissolved, with difficulty, in a mixture of nitric and hydrochloric acids.[1] In this way a platinum compound is formed from which a loosely aggregated "spongy" form of platinum metal can be precipitated.

Shortly before 1800, the English chemist and physician William Hyde Wollaston (1766–1828) worked out a method for putting spongy platinum under heat and pressure in order to convert it into a malleable form that could be hammered into small crucibles and other laboratory ware. Such platinumware was much in demand and, since Wollaston kept the process secret and there were no independent discoverers for nearly thirty years, he grew rich. In 1828, shortly before he died, he revealed his method, but just about that time an even better method was worked out in Russia.

Although platinum was first obtained from Central and South America, the first real mines were developed in the Russian Urals. Between 1828 and 1845, Russia made use of platinum coins. (There is even a story that before that time, some Russian counterfeiters, happening to come across some platinum, made counterfeit coins with platinum re-

---

[1] The mixture is called *aqua regia*, Latin for "royal water," because it dissolves gold, the noble metal, although neither acid will do so by itself—and it dissolves platinum, too, though more slowly.

placing the silver. The only case, that was, of fake coins being better than the real thing.)

Why was platinum so in demand for laboratory ware? Since it reacted even less than gold and was therefore nobler than gold, laboratory equipment made of platinum could be counted on to remain untouched by air, by water or by the chemicals with which it came in contact.

What's more, platinum had a melting point of 1773° C, even higher than that of iron. This meant that platinum-ware could be heated white-hot without damage.

Platinum is denser than gold, too. On the basis of lead's density set equal to 1, gold might be 1.7, but platinum is 1.9.

Finally, it is just as rare as gold is in the crust of the earth.

In that case, if platinum is less reactive, higher-melting, denser and just as rare as gold, isn't it better in every way?

No, it isn't. I've left out one of the characteristics that make gold what it is—beauty. Neither platinum nor any other metal ever discovered has the warm yellowness of gold, and none is anywhere near as beautiful.[2] Platinum can have all the nobility and density and high-meltingness and rareness you can give it, and can even be more expensive than gold, but it will never have gold's beauty, or be as cherished and desired as gold is.

Platinum is not the only metal that is nobler than gold. It is one of three very closely allied metals.

In 1803, an English chemist named Smithson Tennant (1761–1815) noticed that when he dissolved platinum in aqua regia, a black powder was left over that had a metallic

---

[2] There are copper-zinc alloys ("brass") that are gold in color, but they'll develop a greenish rust given the least excuse and that rather spoils things.

luster. It seemed to him that the platinum he had been working with was not pure and that it contained minor admixtures of other metals.

Platinum, however, was the most difficult of all known metals to force into chemical reaction. If there were a metal or metals that were dissolving in aqua regia more slowly than platinum was, those metals had to be hitherto unknown.

Tennant studied the residues carefully, forcing them into solution with considerable trouble, and was able to divide them into two fractions with different properties. One of them formed chemical compounds of a series of different colors, and he therefore named it "iridium," from the Greek word for the rainbow. The other formed an oxide with a foul smell (and very poisonous, too, but Tennant didn't make enough to die of it) and so he called it "osmium," from a Greek word for "smell."

Chemically, iridium and osmium are so like platinum that geological processes throw them together. Wherever platinum is concentrated, iridium and osmium are concentrated, too, so that one always recovers a triple alloy. However, iridium and osmium are only a fifth as common as platinum (or gold) is in the earth's crust, so that the mixture is always chiefly platinum.

Iridium and osmium are, in fact, among the rarest metals in the earth's crust.

Individually, they are like platinum, only more so. Both iridium and osmium are even nobler than platinum, even more reluctant to combine with other compounds. Iridium is, in fact, the noblest metal of them all.

Both are denser than platinum since on the lead-equals-1 basis: iridium is 1.98 and osmium is 1.99. Osmium is, in fact, the densest normal substance known.

Both are higher-melting than platinum. Iridium melts at 2454° C and osmium melts at 2700° C. Here, however, they set no records. The metals tantalum and tungsten melt at temperatures of 3000° and 3400° C, the latter being the highest-melting metal of them all.[3]

Oddly enough, the earth's crust seems to be deficient in the three "platinum metals" (a term that includes osmium and iridium). For every 5 atoms of gold in the earth's crust, there are 5 atoms of platinum, 1 atom of osmium and 1 atom of iridium.

In the universe as a whole, however, it is estimated that for every 5 atoms of gold, there are 80 atoms of platinum, 50 atoms of osmium and 40 atoms of iridium. Why the discrepancy?

There are other atoms in which earth is deficient when compared to the universe as a whole—hydrogen, helium, neon, nitrogen and so on. These do not offer any puzzles. They are elements that are themselves volatile, or that form volatile compounds, so that earth's gravity is not intense enough to hold them.

Platinum, iridium and osmium are, however, not in the least volatile in either elementary or compound form. Why, then, are they missing?

Well, the earth's *crust* is not the earth. The crust can lose elements not only to outer space but also to earth's own interior.

Thus, for every 10,000 silicon atoms in the universe, there are 6000 iron atoms. For every 10,000 silicon atoms in the earth's crust there are only 900 iron atoms. Eighty-five per cent of the iron is gone because it is down in the earth's depths, where there is a liquid metallic core that is chiefly iron. The core also contains a disproportionate share of those metals that tend to dissolve in the iron to a greater extent than to mingle with the crustal rock. The platinum

---

[3] Carbon, a nonmetal, melts at a somewhat higher temperature than even tungsten does, and a compound of tantalum and carbon, tantalum carbide, does even better than either, melting at 3800° C.

metals are apparently readier to dissolve in iron than gold is
and that leaves a deficiency of the former in the crust.

Now let's switch to something else which, at first blush,
seems to have no connection at all with the matter of the
platinum metals. As we shall see, though, science has its
surprises.

There is some value in knowing the rate at which sedimentation takes place in shallow arms of the sea, and how
fast sedimentary rock is formed. That would help us date
fossils; it would help us measure the rate of evolution; it
would help us match up the evolutionary story in different
parts of the world and so on.

We know what the sedimentation rate is here and there
on earth today because we can measure it directly. The
question is, has the rate always been the same or has it been
markedly faster or slower in this or that epoch of geologic
history?

Walter Alvarez of the University of California, together
with several coworkers, had a technique they thought could
be used to establish archaic sedimentation rates. As it
turned out, the technique didn't do that, but while working
with it in rocks dating back to the Cretaceous at Gubbio,
Italy (110 kilometers, or 68 miles, southeast of Florence),
serendipity raised its exciting head. In other words, they
found something they weren't looking for that could be
more valuable than anything they had been expecting to
find.

They were using a neutron-activation technique. This is a
device in which neutrons are fired at a thin slice of rock—
neutrons of an energy which some particular atoms will pick
up with great readiness while other atoms will not. The
atom that does pick up the neutron will be converted into a
known radioactive atom which will break down at a known
rate giving off particular types of radiation. By measuring
the radioactive breakdown the quantity of the particular
neutron-absorbing atom can be measured.

Since radioactive radiations can be measured with great precision, neutron activation techniques can quickly and easily determine the exact quantities of tiny traces of particular atom varieties.

Alvarez tested the delicacy of the technique by setting up the experiment in such a way as to measure the concentration of a particularly rare component of the rocks—iridium. The quantity of iridium in those rocks was, roughly, one atom in every 100 billion. Testing for that iridium atom was something like finding one particular human being in 25 planets each as full of human beings as earth is.

That's a pretty stiff job, but neutron-activation techniques could handle it easily.

And though Alvarez and his associates decided the technique wouldn't solve the particular problem they were tackling, they did come across a narrow region in the rock in which the iridium was 25 times as high as it was everywhere else. That still wasn't much, you understand—one atom in every 4 billion—but, plotted on a graph, that would make an extraordinarily high blip in one specific place in the rock.

How could this happen?

It could be that for some reason, over a relatively short period of time, the seas teemed with iridium (relatively speaking), and that more of it settled out than ordinarily did; or else that the seas had the normal amount of iridium but, for some reason, it settled out 25 times as fast as usual, while other atoms (or at least the common ones) were still settling out at their ordinary rates.

A selectively rapid settling seemed beyond the bounds of possibility so it would seem we are stuck with supposing the presence of abnormally high concentrations of iridium in the sea. If so, where could it come from?

Could there have been some nearby supernova that enormously increased the incidence of cosmic rays that fell upon the earth and could these have induced nuclear reactions that, for some reason, increased the iridium content of earth's outermost layers generally at just that one particular epoch in our geologic history?

If so, there should be other indications. The iridium isotopes should not be in their normal ratios since the most likely changes would produce one particular iridium isotope rather than the other. (There are two stable iridium isotopes.) In addition, there might well be other elements that would be increased in quantity, such as the radioactive isotope plutonium-244 and its decay products. Alvarez ran some quick tests in that direction and his preliminary results seemed negative.

That weakened the likelihood of a supernova as an explanation.

Is it possible, then, that matter from the outside universe was brought to earth bodily? Such matter could be considerably richer in iridium than earth's crust was and this could lead to a temporary 25-fold jump.

The obvious source of such matter would be a meteorite— a huge nickel-iron meteorite, quite like earth's central core in chemical makeup and therefore richer in iridium than earth's crust is. Perhaps it smashed into the Gubbio region and left its mark in the iridium increase.

It is hard, however, to believe that a catastrophic collision would not have left some physical signs in the form of crushed rock, distorted strata, lumps of meteoric iron, and so on. Perhaps the meteorite hypothesis can have its shortcomings ironed out, but I rather think that it is a low-probability explanation.

What else? If not a meteorite, what other form of matter could reach earth?

What about solar material? Suppose at some stage in past history, the sun hiccuped for some reason and had a very mild explosion. Until very recently, this would have seemed most unlikely, but in just the last few years, our studies of the sun have been shaking our faith in it as a steady and reliable furnace. The Maunder minima (see Chapter 1) and the missing neutrinos (see Chapter 2) have worried us a bit. We're somewhat readier to believe in a solar hiccup now than we would have been a decade ago.

Such a slight explosion might have amounted to nothing at all on the solar scale; an insignificant fraction of the solar mass may have blown loose and gone drifting off into space. Some of this finally reached the earth and settled through its atmosphere and ocean into the sedimentary rock, where it mixed with the native material. Since the solar matter would have been richer in iridium than earthly crustal material would be, that would account for the iridium-rich region.

After the explosion was over, the sun would settle down to its accustomed behavior, not measurably different from what it had been before. The solar material on earth would eventually all settle out and the earth would go on as before, too. What's more, the short period of settling of solar material would not be a terrific smashing blow, as of a meteorite. It would be a gentle downward drift. If it weren't for the blip in the iridium, we would never know.

And yet . . . That slight explosion on the sun must have multiplied the amount of heat delivered to the earth. The soft drift of matter must have been accompanied by a most harsh rise in temperature, which may have been only momentary on the geologic time scale but which may have lasted days (or weeks or years) on the scale of life upon the earth.

Such an explosion would have wreaked havoc with life on earth—if it had happened.

Can we argue, then, that since no such havoc seems to have taken place, that the explosion couldn't have happened?

Let us ask first just when this iridium blip took place. According to Alvarez's dating procedures, it happened 65 million years ago at the end of the Cretaceous, and it was precisely at the end of the Cretaceous that the Great Dying took place (see "The Dying Lizards," in *The Solar System and Back*, Doubleday, 1970—an essay in which I discussed a supernova as the possible cause).

Sixty-five million years ago, over a relatively short period of time all the giant reptiles died out, all the ammonites, and so on. It is estimated that up to 75 per cent of all the species living on earth at that time were suddenly wiped out for some unknown reason.

Nor can we assume that the remaining 25 per cent were untouched. It may be that, let us say, 95 per cent of all individual animals were killed, and that the larger ones, who reproduced at a slow rate and were reduced to an unusually small number, could not recover but died out. The smaller ones, who survived in larger absolute numbers and who were more fecund managed to hang on—but just barely.

What it amounts to is this:

About 65 million years ago, earth may have been nearly sterilized, the life upon it nearly wiped out—on the basis of the fossil record.

About 65 million years ago, earth may have suffered a solar accident that would have been capable of nearly sterilizing it—on the basis of the iridium blip.

Can this convergence of two entirely different pieces of evidence be a coincidence?

Of course, it is hard to pin too much on this preliminary work by the Alvarez team, and they make no claim that their speculations of possible astronomical catastrophe are more than speculations. I myself would like to see a thoroughgoing analysis of 65-million-year-old rocks in many places on earth, for a solar explosion would have affected the entire surface, it seems to me. It should also have resulted in raised values for some elements other than iridium, too.

Perhaps the suggestion will turn out to be an utterly false alarm on closer examination. If so, I will confess to feeling relieved, for it is a grisly event that seems to be indicated—

chiefly because, if it happened once, it could happen again, and, perhaps, without warning.

Note

The article above was written in August 1979. Since then, things have progressed rapidly. The supernova hypothesis has lost favor and it can be dismissed (barring further evidence).

Instead, the meteorite hypothesis has gained favor. The iridium blip appears in various parts of the earth and seems to be a global feature. Therefore the meteor can't be just a meteor but must have been a sizable asteroid—10 kilometers across—that kicked so much dust and ash into the stratosphere as to block perceptible sunlight from reaching the earth for *three years.*

Such a long wintry night, if it happened, must have killed off plant life except in such form as could have survived to the end—seeds, spores, roots and so on. All animal life larger than that of a medium-sized mammal must have died—every last dinosaur was dead when the three years were up. Those that survived were the small ones that could live on plant remnants or on frozen animal corpses.

In any case, there will be a full-sized essay on the subject in my next collection. Watch for it!

THE STARS

4.

How Little?

My beautiful blue-eyed, blond-haired daughter is planning to begin graduate courses in psychiatric social work pretty soon and I was on the phone discussing the financial situation with her.

Since she is the apple of my eye and since I am comfortably solvent, no problem arose that would involve economizing and corner-cutting, and the two of us were getting along swimmingly.

And then a nasty little thought occurred to me. Robyn seems to me to be just as fond of me in a daughterly way as I am of her in a fatherly way, but then I have never had to put that fondness under any serious strain by putting her on short rations.

We had not been talking long before I began to feel uneasy, and finally I felt I *had* to know.

"Robyn," I said uncertainly, "would you love me if I were poor?"

She didn't hesitate a moment. "Sure, Dad," she said, matter-of-factly. "Even if you were poor, you'd still be crazy, wouldn't you?"

It's nice to know that I am loved for a characteristic I will never lose.

I *am* crazy, after all, and always have been, and not only in the sense that I have an unpredictable and irreverent sense of humor, which is what Robyn means (I think). I am also crazy in that I make a serious and thoroughly useless attempt to keep up with human knowledge, and feel chagrined when I find I haven't succeeded—which is every day.

For instance—

Years ago, when I first read about the white dwarf companion of Sirius (which is properly termed Sirius B) I discovered that its diameter had been found to be just about equal to that of the planet Uranus, which is 46,500 kilometers (29,000 miles), even though its mass was fully equal to that of the sun. I filed that item away in the capacious grab bag I call my memory, and retrieved it instantly whenever I needed it.

For years, nay, decades, I kept repeating that Sirius B had the diameter of Uranus. I even did so in my book on black holes, *The Collapsing Universe* (Walker, 1977) and in my essay "The Dark Companion," included in *Quasar, Quasar, Burning Bright* (Doubleday, 1978).

The trouble is that the figure I kept giving for the diameter of Sirius B is wrong and has been known to be wrong for a long time now. As one reader said to me (with an almost audible sigh emerging from the paper), the figure I offered was an interesting historical item, but nothing more.

I just hadn't kept up with the advance of knowledge.

Now I have the 1979 figures (which I hope will stay put

for a while) and I will set the record straight. We will consider how little Sirius B really is and how little (alas) I really knew about it.

The diameter of the sun is 1.392×10^{11} centimeters and the diameter of Sirius B is equal to 0.008 times that, or 1.11×10^9 centimeters. If we turn that into more familiar units, then the diameter of Sirius B is equal to 11,100 kilometers, or 6900 miles.

Suppose we compare the diameter of Sirius B to earth and to its two nearest planetary neighbors. In that case, we find:

DIAMETER

	kilometers	miles	earth = 1
EARTH	12,756	7,928	1.00
VENUS	12,112	7,528	0.95
SIRIUS B	11,100	6,900	0.87
MARS	6,800	4,230	0.53

If the question we are asking concerning Sirius B, then, is, how little? the answer is, very little.

Sirius B is smaller in size than either earth or Venus, though it is considerably larger than Mars.

The surface area of Sirius B is equal to 387,000,000 square kilometers (150,000,000 square miles). That is 0.76 that of the surface area of earth. The surface area of Sirius B is about equal to that of earth's oceans.

As for the volume of Sirius B, that is equal to 0.66, or only ⅔ that of earth.

How little? The diameter of Sirius B is only one quarter of what I have been claiming for it all these years and its volume is only one eightieth.

Next, what about the density of Sirius B?

The density of any object is its mass divided by its volume, and the mass of Sirius B, at least, hasn't changed. It's just what I always thought it was—about 1.05 times the mass of our sun. Since the mass of the sun is 1.989×10^{33} grams, which is 332,600 times the earth's mass of 5.98×10^{27} grams, it follows that the mass of Sirius B is equal to $332,600 \times 1.05$, or just under 350,000 times the mass of the earth.

Since the mass of Sirius B is 350,000 times the mass of the earth and since the volume of Sirius B is 0.66 times that of the earth, then the density of Sirius B is 350,000/0.66, or 530,000 times the earth's density.

Earth's average density is equal to 5.52 grams per cubic centimeter. Sirius B's average density is therefore equal to $530,000 \times 5.52$, or 2,900,000 grams per cubic centimeter.

This means that if we imagine an American twenty-five-cent coin (which I estimate to be about ⅔ of a cubic centimeter in volume) to be made up of matter like that in Sirius B, it would weigh about 2.1 tons.

Sirius B does not have the same density all the way through, of course. It is less dense near its surface and grows denser as we imagine ourselves going deeper into its substance until it is most dense at the core. It is estimated that the density of Sirius B at its center is 33,000,000 grams per cubic centimeter. If we imagine a twenty-five-cent piece made of centrally dense Sirius B material, it would weigh about 24.3 tons.

Surface gravity, next.

The gravitational pull of one body on another is directly proportional to the product of the masses and inversely proportional to the square of the distance between the centers of gravity of the two bodies.

If we consider the pull of earth on an object on its surface, then $g = km'm/r^2$, where g is the gravitational pull of earth on the object, k is the gravitational constant, m' is the mass of the object, m is the mass of the earth, and r is the distance between the center of the earth and the center of the object on its surface, this distance being equal to the radius of the earth.

If we next consider the pull of Sirius B on the same object on its surface, then $G = km'M/R^2$, where G is the gravitational pull of Sirius B on the object, k is still the gravitational constant, m' is still the mass of the object, M is the mass of Sirius B, and R is the radius of Sirius B.

To determine how much stronger the surface gravity of Sirius B is than earth's is, we divide the equation for Sirius B by the one for earth, like this:

$$G/g = \frac{km'M/R^2}{km'm/r^2}$$

When we do this, we see that the gravitational constant and the mass of the object on the surface cancels. We get:

$$G/g = \frac{M/R^2}{m/r^2} = Mr^2/mR^2$$

Suppose next that we take earth's mass to be equal to 1 and its radius to be equal to 1. In that case, with $m = 1$ and $r = 1$, we have:

$$G/g = M(1)^2/1R^2 = M/R^2$$

The next step is to get the values for M and R, but in order to keep the equation consistent, we have to get them in earth-mass units and earth-radius units. That was what we used for m and r. Since we know that Sirius B's mass is 350,000 times that of earth and its radius is 0.87 times that of earth, then:

$$G/g = 350,000/(0.87)^2 = 462,000$$

In short, if we imagine an object existing on Sirius B's surface, it would weigh 462,000 times more on Sirius B than it would on earth.

For instance, I weigh 75.5 kilograms but if I imagined

myself on Sirius B, I would weigh just under 35,000,000 kilograms (38,000 tons).

The luminosity of Sirius B—the total amount of light that it gives off—is a direct observation and it doesn't change as our knowledge of Sirius B's dimensions changes.

The luminosity of Sirius B is 0.03 times that of the sun, so that if we imagined Sirius B in place of our sun, we would receive only $\frac{1}{33}$ the light and heat that we now get.

That sounds reasonable considering the fact that Sirius B is an object much smaller than the sun. It is not quite reasonable, though, since Sirius B is so small that on the basis of size alone it could not give as much light and heat as it does.

If two objects are at the same distance from us and are at the same temperature, then the amount of heat we would get from each is proportional to the apparent surface area of each.

For instance, if the sun happened to have two times its present diameter and were at the same distance and temperature, it would present 2×2, or four times, the surface area in the sky, and would deliver four times as much heat and light as it now does. If the sun were three times the diameter it now is and were at the same distance and temperature, it would have 3×3, or nine times, the apparent surface area and would deliver nine times as much heat and light.

It works just as well in the other direction, too. If the sun were one half its present diameter, then, at the same distance and temperature, it would have $\frac{1}{2} \times \frac{1}{2}$, or $\frac{1}{4}$ the apparent surface area and would deliver $\frac{1}{4}$ the light and heat.

If the sun, then, had a diameter 0.173 times its present diameter and were at the same distance and temperature it now is, it would present a surface area and luminosity 0.03 of what it now has. A diameter of 0.173 times its present diameter would, however, amount to 0.173 × 1,392,000, or 240,800 kilometers (150,000 miles).[1]

This small sun, with 0.03 times the surface area of the real sun, is actually much larger than Sirius B. Sirius B has a diameter only 0.008 times that of the sun and a surface area only 0.000064 times that of the sun. With that tiny surface area it still delivers 0.03 times the light and heat of the sun.

In order to account for this discrepancy, we have to suppose that every square centimeter of Sirius B's surface radiates 0.03/0.000064, or about 470 times as much light as every square centimeter of the sun's surface.

The only way that can be is for Sirius B to have a much higher surface temperature than the sun does. This is possible, despite the small size of Sirius B, because it is *not* a main-sequence star. It is a white dwarf star and the rules are different for white dwarfs.

Whereas the surface temperature of the sun is 5600° K (9550° F), the surface temperature of Sirius B is something like 27,000° K (48,600° F) or just about five times as high.

If we were close enough to Sirius B to have its globe seem as large to us as that of our sun now does, Sirius B would be an intensely blue-white object that would broil us to death with heat and fry us to death with ultraviolet light.

Sirius B may be small but it's nothing to fool with.

Of course, to have Sirius B appear as large as the sun would mean that we would have to be fairly close to it. We would have to be only 1,180,000 kilometers (733,000 miles) away from it, and that is only three times the distance from the earth to the moon.

Let us imagine instead that Sirius B existed in place of the sun and was precisely at the sun's present distance.

---

[1] To be sure, a sun that size couldn't possibly be at the same temperature as the sun if it were a main-sequence star (that is, a normal star such as at least 99 per cent of the stars we observe are). However, we're just supposing.

We would then get only 0.03 times the light and heat we now get so that the earth would freeze solid—but let us imagine that, through all the permutations I will suggest, we represent observers on earth who are immune to environmental change.

Since Sirius B has a mass 1.05 times that of our sun, its gravitational pull on the earth would be that much stronger and the earth would revolve somewhat more quickly. The year would be only 356.5 days long.

Sirius B in the position of our sun, would have an apparent diameter of only 15 seconds of arc; that is, it would appear the size that the planet Saturn appears to be when it is farthest from us. Sirius B, therefore, would be visible as a star rather than as a solar globe.

It would, however, be an enormously bright star. It would have a magnitude of −23.8, which would make it 14,000 times as bright as the full moon appears to us now.

While the light of Sirius B, under the conditions described, would be substantially dimmer than the light of our sun, the little star would pose a problem—at least if we were observing it with the kind of eyes we now have. It would be dangerous to look at Sirius B. For all its dimmer total radiation, Sirius B would be sending out far more ultraviolet than our sun does, and I suspect that eyes like our own would be blinded if we unwarily caught a good look at it.

But suppose, then, that earth were *not* circling Sirius B, but were circling the sun exactly as it is doing. And suppose that Sirius B were the companion of our sun as it is, in actual fact, the companion of Sirius A. If we saw Sirius B not in the place of our sun, but as our sun's companion, revolving about the sun in the plane of the planetary orbits, what would it look like?

Sirius B and Sirius A circle a common center of gravity with an orbital period (for each) of 49.94 years. This, however, takes place under the gravitational lash of the combined masses of the two stars. Sirius A, the bright normal star that is the jewel of our heavens, has a mass equal to 2.5 times that of our sun, so that the combined mass of Sirius A and Sirius B is 3.55 times that of our sun.

If Sirius B were imagined to be circling our sun instead, in precisely the same orbit that it circles Sirius A, then its orbital period would lengthen at once. The combined mass of the sun and Sirius B is only 2.05 times that of our sun alone, so that the gravitational pull that would drive the objects in their orbits would be correspondingly less than for the combination of Sirius A and Sirius B.

Sirius B and the sun would be circling a common center of gravity (located about midway between them) with an orbital period of 65.72 years.

The mean distance of Sirius B from Sirius A is 3 billion kilometers (1.9 billion miles) and if this were true for the Sirius B and sun combination, it would mean that Sirius B would be somewhat more distant from the sun than the planet Neptune is.

Sirius B and the sun, however, would not maintain a constant distance, for Sirius B and Sirius A follow orbits that, in actual fact, are markedly elliptical, and we must suppose the same for Sirius B and the sun.

The orbital eccentricity of Sirius B's orbit relative to Sirius A and, therefore, relative to the sun in our imagination, is 0.575. That means that the distance between itself and the sun would vary from as little as 1.28 billion kilometers (800 million miles) to as much as 4.72 billion kilometers (3 billion miles).

In terms of our solar system, then, Sirius B would sometimes be closer to the sun than Saturn is and, at the opposite end of its orbit, recede to slightly farther than Pluto at its most distant.

Under those conditions, the sun's outer planets would scarcely be moving in stable orbits and we can assume they wouldn't exist. The inner solar system, including earth, would not be seriously affected by Sirius B, however, and we would circle the sun as always.

In that case, what would Sirius B look like in the sky?

If it looks like a star, with no visible disk, even when it is in place of our sun, it would certainly look like a mere star at the distance of Saturn. It would be correspondingly dimmer, too, naturally.

When Sirius B, as companion to the sun, was closest to the sun, and if we then happened to be located in that portion of our orbit that would be between the sun and Sirius B, we would be 1.13 billion kilometers (707 million miles) from Sirius B. It would then have a magnitude of −19.4 and be only 1/1000 as bright as the sun. Still, 1/1000 is a respectable fraction, actually, for Sirius B would then be 465 times as bright as the full moon is now.

Even then, Sirius B would be an uncomfortable thing to look at, I should think. At its high temperature, as much ultraviolet light might be reaching us from Sirius B at the distance of Saturn as from the sun at its much closer distance.

It strikes me that our moon might present an interesting appearance in such a system—possibly a three-tone look. If the earth, the moon, the sun and Sirius B were properly oriented, we might, for instance, see a rather thin crescent facing the west, another, much dimmer crescent facing the east, and darkness in between. The moon, as it circled the earth, would undergo a double phase change of marvelous intricacy.

As the earth went around the sun, Sirius B would appear to move in the sky relative to the sun, remaining in the night sky for different periods of time, just as any of the planets do now. There would be times when Sirius B would rise at sunset and set at sunrise and be visible in the sky all night through. In that case, the night would not be truly dark. It would be dimly twilit throughout.

The pattern of day, night, and "companion light" would vary through the course of the year.

When Sirius B shone in the sky during some of the daylight hours, it would shine as a visible point of light and everything would have a very faint shadow in addition to its normal shadow, the two being at changing angles to each other in the course of a year.

This is all when Sirius B is nearest the sun. From year to year, though, it would get fainter as it moved farther and farther from the sun. So would the companion light and the second shadow. Finally, Sirius B would reach its farthest point, nearly thirty-three years after it had been at its nearest point.

At its farthest point, Sirius B would have a magnitude of only −16 and would be only twenty-three times as bright as the full moon is now. Thereafter, it would begin to brighten again.

Next to the rising and setting of the sun and to the phases of the moon, this slow brightening and dimming of Sirius B would be the most remarkable cycle in the sky, and it seems to me that the period of the cycle would be given enormous importance.

The slow cycle of Sirius B would, after all, almost match the normal lifetime of a human being, and no doubt primitive people would imagine Sirius B to be matching the beat of human life. Think of the fun astrologers would have had with *that*, and thank heaven we are spared it.

Sirius B was not always a white dwarf, of course. Once upon a time it was a main-sequence star like the sun. We can suppose that it was not much more massive then than it is now, and that it was not massive enough to undergo a supernova explosion once its hydrogen fuel was consumed. It merely expanded to a red giant and then collapsed non-catastrophically.

As an ordinary star (following the same orbit we imagined Sirius B to have as our sun's companion), Sirius B would have been perhaps thirty-five times brighter at every stage than it would be as a white dwarf. At its closest approach, it would be 1/30 as bright as our sun and over

16,000 times as bright as the full moon. Even at its farthest recession, it would be 800 times as bright as the full moon now is.

Nor would Sirius B appear to have a solar globe most of the time, even as a normal star. At its closest, though, it would be nearly 6 minutes of arc across and would seem like a tiny little circle of light.

And then the time would come when enough of the hydrogen fuel would have been lost for helium burning to begin at the center of Sirius B. That would mean it would begin expanding in size, and its surface would cool and redden as a result.

It would be a fascinating change, as Sirius B, which would be by far the brightest object in our sky, next to the sun, would slowly grow and redden.

The process might take many thousands of years and the change, I daresay, would not be visible in the lifetime of a particular person. However, the scientific records over the course of generations would make it obvious that Sirius B was growing and reddening. Finally, the growth would slow and stop and the red orb would reach a maximum size.

We might guess that its diameter would then be something like 200 million kilometers (124 million miles).

In that case, when Sirius B was farthest from the sun, we would see it in the sky as a circle of red light with a diameter of nearly 1.4°. It would be 2.56 times as wide as the sun appears to us now and it would have 6.57 times the area. It would, however, have so cool a surface that it would deliver considerably less heat than the sun does.

At its closest, the Sirius B red giant would have a diameter more than 4 times what it had at its farthest. It would then have a little more than 25 times the surface area of the sun.

Under such circumstances we would have a pattern of white light when the sun was in the sky; orange light when sun and Sirius B were together in the sky; red light when only Sirius B was in the sky; and darkness when neither was in the sky. When both were in the sky, there would be red shadows and white shadows set at angles, turning black where they overlapped near the object that cast them.

The red giant would remain at its peak for a long period of time—a million years, perhaps—and then the time would come when it would collapse suddenly, perhaps in a matter of hours. It would leave behind it a ring of gas, marking its outer limits (thus forming a "planetary nebula") and at the center there would suddenly be a white dwarf. The ring of gas would expand and grow thinner, engulf earth and sun and gradually vanish. Only the white dwarf would be left and, we hope, photographic records of the red giant, or future generations would scarcely believe in its existence.[2]

Sirius B may not have behaved this way in actual fact. It might have been a much more massive star when it was on the main sequence. Then, as it expanded to a red giant, matter from it may have been spilled over into Sirius A. This may have saved Sirius B from exploding violently, but it also increased the mass and brightness of Sirius A and shortened its ultimate lifetime.

It is even possible human beings witnessed the change. I have heard that a number of ancient astronomers described Sirius as being red in color and, if so, they could scarcely have been mistaken about it. Neither can present astronomers be mistaken in seeing Sirius as blue-white.

Could it be that the ancients were watching not the Sirius A we see, but Sirius B as a red giant while it was bleeding matter over to a relatively dim Sirius A?

Then at some time in the early Middle Ages, when astronomy was at a low ebb and the sudden change went unnoticed, Sirius B may have collapsed and become too dim to be visible with the unaided eye, leaving behind the suddenly enhanced blue-white sparkle of Sirius A.

We'll return to this matter in the next chapter.

---

[2] Mind you, I am ignoring the fact that the formation of the red giant would probably wipe out life on earth.

5.

Siriusly Speaking

I suppose it's no secret that I was a child prodigy. (It can't be a secret, says my good friend Lester del Rey, because I talk about it all the time.)

Of course, I know many people who have been child prodigies, but, as far as I know, I was the only one to enjoy it. I just loved being smarter than the next kid.

What's more, I wasn't bored in class because there was always the teacher to bedevil. I wasn't lonely through inability to relate to children my age because, aside from being a prodigy, I was adequately childish. I wasn't ostracized by my peers for the crime of being bright because I was also a disciplinary problem in class, which made them all decide I was a decent sort.

In fact, I love having been a child prodigy so much I hate to let go. Being an adult prodigy always seems pale in comparison, but at my age, I think that in a few years I'm going to have to settle for it.

Anyway, I've been leafing through *The Science Fiction Encyclopedia* (Doubleday, 1979) and it suddenly occurred to me that it tells me just how old all my good science-fiction-writing buddies are—so that I can make invidious comparisons.

For instance, a decade ago the Science Fiction Writers of America voted on the best short stories of all time, with the rules requiring that only one story be chosen from any one author. As a result, the "best" stories of each of twenty-six different authors were chosen and placed in an anthology.

I have just gone over the list of stories in that anthology and, with the help of *The Science Fiction Encyclopedia*, have worked out the age of the author at the time each of the stories was published.

The senior author to achieve the list was Murray Leinster, who published his classic "First Contact" when he was forty-nine. The junior author to achieve the list (are you ready?) was Isaac Asimov, who published his classic "Nightfall" at the age of twenty-one.

I'm not through. Which story of the twenty-six finished in first place (even though I didn't vote for it myself)? You're right. It was "Nightfall."

Now I have always maintained that "Nightfall" is *not* the best story I've ever written, let alone the best ever written by anyone (whatever the votes say). I am willing to say this much, however, in my child-prodigy pride. It's the best science-fiction story of *any* length ever written by *any* person who had not yet attained his twenty-second birthday.

And why am I telling you all this? Because in an essay I recently wrote to correct an insufficiency of knowledge displayed in an earlier essay, I managed to commit another and worse insufficiency of knowledge, and now I have to correct *that*. Feeling stupid, I had to do something to cheer myself up.

I've done so and now I can begin:

Imagine a sphere in space with a radius of 8.8 light-years and the sun at its center. The volume of that sphere is 2850 cubic light-years.

That's a big sphere. The sun, at the center, giant globe though it is, takes up only 5×10^{-24} (five trillion trillionths) of that volume.

Imagine a smaller sphere with the sun at the center; one with a radius of 7.4 billion kilometers (4.6 billion miles). That would be large enough to include even the orbit of Pluto, and all the solar system, in fact, but for some far-flung comets. This solar-system sphere would still be only 7×10^{-11} (70 trillionths) of the volume of the larger sphere. In comparison with the 8.8-light-year-radius sphere, the entire solar system shrinks to a dot.

Yet that huge sphere becomes small if viewed in another way. A sphere with a radius of 8.8 light-years takes up only about 1.4×10^{-10} (1/7 of a billionth) of the volume of the Milky Way galaxy.

This is roughly the area of a square city block as compared with that of the United States. We are perfectly correct, then, in considering the 8.8-light-year-radius sphere as representing our immediate stellar neighborhood; our own city block in the midst of a large nation.

The question that then arises is, how many stars are there in our immediate neighborhood? How many stars live on our block?

The answer is, counting our sun, exactly nine.

Working outward from the sun, we come to the Alpha Centauri system of three stars which are, in order of decreasing brightness: Alpha Centauri A, Alpha Centauri B and Alpha Centauri C. The first two are so close to each other that they may be considered to be at the same distance from ourselves.

Alpha Centauri C, however, is at an immense distance from the other two. It's about 1.6 trillion kilometers (a trillion miles) from the center of gravity of the A/B system and takes 1.3 million years to make a circuit about it. Right now, it is in that part of its orbit that puts it between the A/B system and ourselves so that it is measurably closer to us than they are. Alpha Centauri C is, indeed, the closest known star to us and it is about 4.27 light-years away.

Alpha Centauri C is, however, a small red dwarf, a bit more than 1/20,000 as bright as our sun. Despite its proximity (by stellar standards), its visual magnitude is 11.05, which makes it far too dim to see with the naked eye. It would have to be a hundred times as bright as it is to show up as a barely visible star in the sky.

Alpha Centauri A and Alpha Centauri B are 4.37 light-years away. Of the two, Alpha Centauri A is virtually the twin of our sun—the same mass, the same temperature, the same luminosity.

Alpha Centauri B is a substantially smaller star. It has a diameter of 973,000 kilometers (605,000 miles) as compared to 1,390,000 kilometers (864,000 miles) for Alpha Centauri A or the sun. Alpha Centauri B is only about 0.28 times as luminous as either Alpha Centauri A or the sun.

Either Alpha Centauri A or Alpha Centauri B, if shining alone in the sky, would be a "first-magnitude" star. Alpha Centauri A has a visual magnitude of −0.01 and Alpha Centauri B one of 1.33.

We do not see them separately, however. They are so close to each other that, to the unaided eye, they seem like a single dot of light, with a visual magnitude of −0.1. This makes it the third brightest star in the sky.

Moving out from the Alpha Centauri system, we encounter three single stars: Barnard's star, Wolf 359, and HD 95735, at distances of, respectively, 5.9, 7.6, and 8.1 light-years. These are all red dwarfs that are invisible to the naked eye.

Of these three, the brightest is HD 95735, which has a hundred times the luminosity of Alpha Centauri C, and a visual magnitude of 7.5. Were it but two and a half times more luminous than it is, it would be just visible to sharp eyes on a very clear moonless night.

Barnard's star is about midway in luminosity between Alpha Centauri C and HD 95735, and has a visual magnitude of 9.54.

The dimmest of all the stars on our block is Wolf 359. It is only about a third the luminosity of Alpha Centauri C and has a visual magnitude of 13.53, despite its relative closeness to ourselves.

That gives us seven of the nine stars. For the last two, we must move out to a distance of 8.65 light-years, almost at the edge of the block and there we will find an oddly assorted pair of stars, Sirius A and Sirius B.

Concerning Sirius B, I have spoken in "How Little?" (Chapter 4). It is a white dwarf star with the mass of the sun and a volume just under that of the earth. Its visual magnitude is 8.68, which places it between HD 95735 and Barnard's star in apparent brightness. It is even further at a disadvantage for it is drowned out in the resplendent light of its companion, Sirius A.

Sirius A is the aristocrat of the block. It is both larger and hotter than the sun. In diameter it is 2.5 million kilometers (1.55 million miles) across, which makes it very nearly 1.8 times as wide as the sun and 5.8 times as voluminous. It has 2.5 times the mass of the sun and it has a surface temperature of 10,000° C as compared to the sun's 6,000° C. All told, it is 23 times as luminous as the sun.[1]

---

[1] In all these properties, it holds no records. There are stars far more more voluminous, and massive, and hotter, and more luminous than Sirius—but not on our block.

As a result of Sirius A's luminosity and its closeness to us, its apparent visual magnitude is −1.45, making it the brightest star in the sky by a considerable margin.[2]

Sirius A (which from now on we need call only Sirius) was bright enough to attract the attention of the ancients. It would have done so in any case, but the Egyptians paid it special attention owing to a peculiar coincidence. Let me explain.

The sun moves in the sky relative to the stars generally, an effect produced by the earth's revolution about the sun. Every twenty-four-hour period, the sun moves 0.986° eastward against the background of the stars so that, at the end of a solar year of 365.2422 days, it has moved 360° eastward and has made a complete circle about the sky.

If we consider some particular star that happens to rise with the sun on some particular day ("heliacal rising," from Greek words meaning "near the sun") the next morning that star will be 0.986° west of the sun and will rise 3.95 minutes earlier than the sun, and the next morning 3.95 minutes earlier still. It will go more and more out of phase until, after six months, the star will be rising as the sun is setting the night before (so to speak). Then, finally, after exactly one year, that star will be rising with the sun again.

The Egyptians noticed that the star Sirius (which they called Sothis) rose with the sun at just about the time the Nile rose in flood, and they felt this to be an important signal from the gods and watched for it.

It was clear that the heliacal rising of Sirius occurred every 365 ¼ days and that helped the Egyptians set up a solar calendar that ran 365 days to the year and that ig-

[2] If Sirius is the brightest and Alpha Centauri A/B is the third brightest, which is the second? That is Canopus with a visual magnitude of −0.73. Canopus is nearly 200 light-years away, however, and achieves its apparent brightness only because it is some 5200 times as luminous as our sun and 225 times as luminous as Sirius A.

nored the phases of the moon—this at a time when other ancient people used the less convenient and more complicated lunar calendar.

To be sure, the Egyptians ignored the fraction of a day and kept the year a steady 365 days, year in and year out. That meant that the calendar lost a quarter day each year. If the heliacal rising of Sirius took place on January 1 in a particular year, it would take place on January 2 four years later, January 3 four more years later and so on. (This could have been avoided if the Egyptians had adopted a leap-year convention. They knew enough to do so, but clung to tradition—just as we cling to the stupidities of our own calendar because of tradition.)

After 1460 years (365 × 4), Sirius would rise on January 1 again, and so 1460 years was called the "Sothic cycle."

Not only did all this mean that Sirius rose with the sun on different days of the year as time went on, but that the Nile flood arrived on different days of the year. Still, the Egyptians could count on the heliacal rising of Sirius as harbinger of the Nile flood whatever the day of the year.

Now imagine Sirius to be rising earlier every morning and catching up to sunrise again. Suppose a day comes when it rises, say, ten minutes after the sun. Naturally it would be invisible in the solar glare by the time it rises.

The next day it would rise four minutes earlier and top the horizon only six minutes after the sun—and would still be invisible. The next day it would rise two minutes after the sun—and would still be invisible.

The next day, however, it would rise two minutes *before* the sun and it would be seen very low on the eastern horizon in the brightening dawn for just a very brief time before the sun's edge peeped above the horizon.

After that, Sirius would rise earlier and earlier and be higher and higher in the sky at sunrise, but the Egyptians would not be interested in it then. It was the heliacal rising that gripped them.

In 3000 B.C., the heliacal rising of Sirius came about three days before the normal time of the beginning of the Nile flood at the Egyptian capital of Memphis. However, what with the precession of the equinoxes and Sirius' own motion, the heliacal rising slowly drifted. By 2000 B.C., the heliacal rising of Sirius came five days *after* the start of the Nile flood and by 1000 B.C., it came twenty-three days after. Sirius no longer served as harbinger, but by then the Egyptian calendar and traditions had been fixed.

Now comes the mystery I mentioned at the end of the previous chapter. A number of the ancients reported Sirius to be a red star though we ourselves see it as pure white. Why should that be?

In "How Little?" I advanced the notion that Sirius B, before it was a white dwarf, had been a red giant and at that time might have drowned out the glitter of Sirius A. Perhaps the collapse from red giant to white dwarf had occurred some fifteen hundred years ago. Sirius B would then have disappeared as a visible object, and would have left Sirius A glowing. Sirius would then have turned from red in ancient times to white in medieval and modern times.

I actually thought this notion was original with me and presented it while simpering with modest pride, and that is the "insufficiency of knowledge" I referred to in the introduction to this essay.

Fortunately, I have readers who are knowledgeable in any field in which I expose my ignorance and they write to me at once. In this case, it was Dr. Charles F. Richter (of the famous Richter scale for measuring the intensity of earthquakes) who wrote to tell me of many earlier suggestions of this precise theory.[3]

Worse yet, the theory is not a tenable one, however often it is and has been advanced.

[3] I have used his letter, and an article, "Sirius Enigmas," by Kenneth Brecher, in a book entitled *Astronomy of the Ancients* (MIT Press, 1979), for what follows.

When a red giant shrinks to a white dwarf, the result is a "planetary nebula," and the white dwarf is surrounded by a haze of gas. Slowly the haze of gas expands and thins out and eventually is no longer visible; but this takes time.

If the collapse of Sirius B had taken place fifteen hundred years ago, there would still have remained visible traces of this haze. Such traces have been searched for and there are no signs of them. Sirius B could not therefore have collapsed one or two thousand years ago; one or two hundred thousand years ago would be more like it.[4]

In that case, how else can we explain the fact that the ancients reported Sirius to be red?

One possibly useful suggestion was first advanced by G. V. Schiaparelli (the Martian-canal Schiaparelli) a century ago.

Consider that the great astronomical event of the year for the early Egyptians was the heliacal rising of Sirius. As the time approached, they must have watched with religious excitement for the first glimpse of Sirius in the eastern horizon in the bright desert dawn.

And when Sirius appeared just above the horizon it looked reddish for the same reason that the setting sun or the rising sun looks reddish. Light scattering is more efficient the shorter the wavelength and takes place more extensively when a great thickness of atmosphere is traversed. The light from the sun, or a star, passes through a greater-than-usual thickness of the atmosphere when at the horizon; the short wavelengths of light are scattered and the long wavelengths of red and orange tend to survive.

Most stars are so dim that when part of their light is scattered, what is left isn't bright enough to make an impression on the relatively insensitive color-vision apparatus of the human eye. Sirius, in fact, is the only star bright enough to look red at the horizon.

---

[4] Therefore, the offending paragraphs in "How Little?" have been omitted in this collection.

It would be natural, then, for the Egyptians to think of Sirius as reddish. To be sure, Sirius is white when it is high in the sky but it is red *when it counts*—at the time of heliacal rising.

We don't have enough of the ancient Egyptian astronomical records to be sure that they classified Sirius as red, but it seems a reasonable hypothesis and the Greeks may have been influenced by this Egyptian view.

Next, consider this:

During the heyday of the Greeks, the heliacal rising of Sirius took place in the second half of July, at the time of maximum summer heat.

There was a natural tendency to imagine that Sirius, which was so unusually bright for a star, delivered a substantial amount of heat to earth and that when this was added to that of the sun all day long, as it was, near the time of the heliacal rising, the earth would suffer especially high temperatures.

Sirius is the brightest star of Canis Major ("Great Dog"— often viewed as the hunting dog of the nearby constellation of Orion, the Hunter). It was because of Sirius' supposed added heat that the hottest weeks of summer were called the "dog days" and, in fact, are *still* so called. Furthermore the added heat of the dog days was supposed to breed pestilences and fevers (very likely, if it hastened the spoiling of food and the activity of parasites).

Here, then, is how Homer, at the start of the fifth book of the *Iliad*, described the glory of Diomedes as Athena helps him arm for battle: "She kindled a flame that blazed steadily from his helmet and shield, like the star that shines brightest of all in late summer after his bath in the ocean. Such was the fire that blazed from the head and shoulder of Diomedes. . . ."

Sirius (the star that shines brightest of all) seems here to be equated with "fire" and "flame." Does that refer to Sirius' ruddiness, as is sometimes suggested? Or does it refer to Sirius' traditional heat? My own feeling is that it refers to the latter.

Then, in the twenty-second book of the *Iliad,* Sirius is referred to again in connection with Achilles' armor: "His armour shone on his breast like the star of harvest whose rays are most bright among many stars in the murky night; they call it Orion's Dog. Most brilliant is that star, but it is a sign of trouble, and brings many fevers for unhappy mankind."

There the baleful aspect of Sirius is emphasized.

Virgil, who carefully copied every aspect of Homer, has Aeneas' armor gleam like that of Diomedes or Achilles in the tenth book of the *Aeneid.* There it says: "Aeneas' helmet blazed. A stream of fire poured from his plumy crest. A golden fount gushed from the great shield-boss.—As on a clear night comets glow with a grim and blood-red gleam, or as the glare of Sirius, the star that brings to frail mortality disease and thirst and rises sicklying heaven with boding light."

Here again, the reference is to Sirius' baleful aspect.

But what about the association of Sirius with the "blood-red gleam" of comets? Doesn't that mean that Sirius is also blood-red?

Not at all! The fact is that comets are *not* blood-red. However, comets are considered baleful omens of disaster, predicting death and destruction to mankind. The feeling is that they portend war, murder, civil turmoil, all sorts of bloodletting violence. Hence, they predict "blood-red" events and are, by poetic compression, described as "blood-red" themselves.

It is a terrible mistake to expect poets to be literal rather than metaphoric and to base sober theory on the expectation of literalness.

To take a modern example, when Emily Dickinson said, "Not one of all the purple host that won the flag today . . ." do you suppose that she was talking necessarily of people with purple skins or purple uniforms?

No! What Dickinson was doing was making a condensed poetic reference to the tradition, as old as the Byzantine Empire, that equates purple with royalty. The winning host were kings of the field, and were therefore purple.

To be blue is to be of a certain color, *or* it is to be in a state of emotional depression.

To be green is to be of a certain color, *or* it is to be young and/or inexperienced.

If you're a "golden boy" you're multitalented, if you're "yellow" you are cowardly and, in either case, your actual skin color may be pink and white, or chocolate brown.

I'm sure it's the same in other languages, and that in every one of them color words have numerous associations that have nothing to do with the color itself except, at best, through distant analogy or poetic association.

Anything dire and baleful is bound to be thought of as bloody, and anything bloody is bound to be thought of as red. It is therefore not surprising at all if people speak of anything that is lowering and threatening as red.

By Virgil's time, it may well have been fashionable to speak of Sirius as red, referring not to its literal color but to the fact that it threatened mankind with misfortune. Thus, Seneca, in A.D. 25, a generation after Virgil, said, "The redness of the Dog Star is deeper, that of Mars milder, that of Jupiter nothing at all." But was he speaking of literal color or of the intensity of misfortune portended, in the astrological sense? I suspect the latter.

If we eliminate all poetic references, we are left with one overriding problem involving Ptolemy, the crowning astronomer of ancient times.

There are five bright stars that visibly show a reddish or orange tint. They are Aldebaran, Antares, Arcturus, Betelgeuse and Pollux. Ptolemy mentions them all as *hipokeros,* a word which may be translated as "reddish" or "yellowish."

But he adds a sixth! Sirius!

How is that possible?

Well, consider first that Ptolemy lived and worked in Egypt and must have been surrounded by Egyptian records and ways of thought, and we already know that Egyptians might naturally think of Sirius as reddish.

Secondly, as pointed out by Kenneth Brecher, the copies

of Ptolemy we now have are certainly not the eighteen-hundred-year-old originals. The oldest copies we have can only be traced back a thousand years and have already been translated from Greek to Arabic and back to Greek. Who knows what errors of copying and translation may have crept in?

In the oldest copy, for instance, we have a summary at the end which says that there are "five red stars"—which is correct. Was Sirius added later in the body of the book by someone who was overly influenced by poetic descriptions or Egyptian tradition, or by sheer accident?

By the tenth century, the Arabian astronomers were listing only five red stars and omitting Sirius—and they must surely have had access to copies of Ptolemy older than any we now possess.

My conclusion? The mysterious redness of Sirius in Greek times is no mystery because it wasn't red, and was never said to be red in any literal sense.

6.

Below the Horizon

The role of a writer is hard, for on every hand he meets up with critics. Some critics are, I suppose, wiser than others, but there are very few who are so wise as to resist the urge to show off.

Critics of science popularization always have the impulse to list every error they can find and trot them out and smile bashfully at this display of their own erudition. Sometimes the errors are egregious and are worth pointing out; sometimes the critic is indulging in nitpicking; and sometimes the critic inadvertently shows himself up.

I've got a review of one of my science collections in my hand right now. Never mind where it appeared and who wrote it—except that the critic is a reputable professional as-

tronomer. The point is that three fourths of the review is a
listing of my errors.

Some of the errors referred to by the critic are well taken
and I'll have to be more careful in the future. Other errors
he listed I found simply irritating.

After all, in writing on science for the public, you must
occasionally cut a corner if you are not to get bogged down
in too much off-target detail. Naturally, you don't want to
cut a corner in such a way as to give a false impression. If
you must simplify, you don't want ever to oversimplify.

But what is the boundary line between "simplify" and
"oversimplify"? There is no scientific formula that will give
you the answer. Each popularizer must come to his own
conclusion with respect to that, and to do so he must con-
sult his own intuition and good sense. While laying no claim
to perfection, you understand, I hope you won't mind my
saying that in this respect my intuition is pretty good.

But to the point . . .

The reviewer says: "Elsewhere he [Isaac Asimov] states,
incorrectly, that 'as seen from the United States . . . Alpha
Centauri . . . is always below the horizon.' It can, in fact,
be seen from the lower part of Florida every night during
the summer months."

(Of course, he can be nitpicked as well. By "the lower
part of Florida" he means the southern part. Apparently he
assumes that the north-is-up convention in modern maps is
a cosmic law. And he doesn't really mean that it is seen
every night; he means every night that the clouds don't interfere. See how easy nitpicking is, Professor Reviewer?)

However, even an irritating review can be useful since I
can now go into the matter of just which stars can be seen
from which points on earth.

To begin with, I will make some simplifications which I will
specify in full, lest I be nitpicked either for having not
made them or for having made them without stating the
fact.

1. We will suppose that the earth is a perfectly smooth body with no surface irregularities whatever. I rather think that it doesn't matter for the purposes of this essay that it is an oblate spheroid, but as long as we're simplifying, let's go all the way. Let's suppose it is a perfect mathematical sphere so that from any point on earth we will see a true, perfectly circular horizon.

2. We will suppose that the atmosphere does not absorb light. We will suppose there are no clouds, no fogs, no mists, no smoke. Every star that is bright enough to see with the naked eye is seen.

3. We will suppose that only the stars exist in the sky. There is no sun to blank out the stars in the daytime. No moon, planets, comets, or any other solar system objects to confuse the issue. Just the stars!

4. We will suppose atmospheric refraction does not exist. In actual fact, refraction tends to make a star appear higher above the horizon than it really is (unless it is directly at zenith) and since this effect is the more pronounced the closer the star is to the horizon, a star which is distinctly below the horizon can actually be seen slightly above. We will ignore this and suppose that light travels from a star to our eye in a perfectly straight line without being affected by either refraction or, for that matter, any gravitational field.

5. Let us suppose that the earth's orientation with respect to the stars is absolutely unchanging. This is not so, of course, for the orientation changes in several ways:

a. The earth's axis shifts with time, so that if we imagine it to be extended to a point in the sky at both ends, each point marks out a slow circle with time. The earth takes nearly twenty-six thousand years to shift so as to describe that circle which is called "the precession of the equinoxes."

b. The earth's axis inclines more to the ecliptic and then less to the ecliptic by a matter of 2.5° in a cycle that is forty-one thousand years long.

c. The position of the North Pole on the earth's surface varies from moment to moment so that it describes an irregular circle that deviates from the average by distances of up to a couple of hundred meters.

d. The land we stand on is slowly moving as the tectonic plates shift.

6. We will assume that the stars are not themselves changing position relative to each other. Of course, all the stars *are* moving, but except for some of the very nearest, these motions are so damped out by huge distances that even our best instruments can scarcely detect any change at all over a lifetime. For the very near stars, where "proper motion" can be measured by astronomers, the motion is still not great enough to be noticeable to the naked eye over a human lifetime.

All these simplifications do not introduce any substantial errors in what is to be my exposition.

Next let us describe the sky with reference to the earth.

To the eye, the sky appears to be a solid sphere that encloses the earth. If we wanted to make a three-dimensional model of the universe we could make a small sphere with the continents and oceans painted upon it. That would be the terrestrial sphere. Around it we could construct a larger concentric sphere (one with the same center as the smaller one) and call it the celestial sphere.[1]

On the celestial sphere we can mark off the stars as we see them in the sky. This ignores the fact that the sky is not really a spherical surface but that it is an endless volume and that the stars are not at the same distance from earth but at wildly different distances. From the standpoint of this essay, however, the markings on the sphere are sufficient.

How do we locate the stars on the celestial sphere?

[1] "Celestial" is from the Latin word for "sky." "Ceiling" comes from the same Latin word.

To begin with, let's extend earth's axis in imagination until it reaches the sky in both directions. The northern end of the axis reaches the sky at the North Celestial Pole and the southern end of the axis would reach it at the South Celestial Pole.

If we were standing precisely at the North Pole, the North Celestial Pole would be at the zenith, directly overhead. The South Celestial Pole would be at the nadir, on the spot on the celestial sphere that is on the other side of the earth directly under our feet. If we were standing precisely at the South Pole, it would be the South Celestial Pole that would be at zenith and the North Celestial Pole that would be at nadir.

On earth, we can draw a circle about the surface in such a way that every point on that circle is exactly halfway between earth's North Pole and South Pole. The circle is the equator, so called because it divides the earth's surface into two equal halves. You can draw a similar circle on the celestial sphere and you will have the celestial equator.

If you are standing anywhere on the equator, then the celestial equator will be a line across the sky starting on the horizon due east, passing through the zenith and ending on the horizon due west.

Just as you mark off the surface of the terrestrial sphere into parallels of latitude and meridians of longitude, so you can mark off the celestial sphere into parallels of celestial latitude and meridians of celestial longitude.

If earth and sky were at rest with respect to each other, every star in the sky would be exactly at zenith with respect to some point on the surface of the earth. The celestial latitude and longitude of that star would be precisely the latitude and longitude of the point on earth's surface over which it stood at zenith.

As a matter of fact, though, the earth turns from west to east, completing one turn with respect to the stars in twenty-three hours and fifty-six minutes.[2] Naturally, to us standing on the earth it seems as though we were motionless and that the sky was turning from east to west in twenty-three hours and fifty-six minutes per turn.

The apparent rotation of the celestial sphere is equal and opposite to that of the real rotation of the earth and it takes place on the same axis. That means that the Celestial North Pole and the Celestial South Pole remain fixed in the sky. All other points in the sky make circles parallel to the celestial equator. That means their celestial latitude does not change with time.

The celestial longitude does change and that means the complication of an accurate clock must be brought in. In this essay, however, we're concerned only with celestial latitude, which is a good break for us.

Celestial latitude is usually referred to as "declination" by astronomers and is marked off as plus and minus from the celestial equator, rather than as north and south. On the terrestrial sphere, for instance, we would speak of latitudes of $40°$ N and $40°$ S, but on the celestial sphere, we speak of declinations of $+40°$ and $-40°$.

Now, then, let's imagine we are standing precisely at the North Pole. The North Celestial Pole is at zenith and, as the celestial sphere turns, it stays there. The entire celestial sphere pivots around it and every point on the sphere describes a circle parallel to the horizon. The celestial equator is exactly at the horizon at all points.

This means that every star in the sphere that is in the north celestial hemisphere, and therefore has a positive declination, remains above the horizon at all times and is

---

[2] It takes another four minutes for the turning earth to catch up with the position of the sun in the sky, since in the interval the sun has moved slightly with respect to the stars. It is with respect to the sun that we measure the length of the day. That makes the day twenty-four hours long.

therefore visible. (Remember we are ignoring sun, clouds, haze, refraction and all other phenomena that would tend to spoil our pretty theoretical picture.)

If every star in the north celestial hemisphere is forever visible, as seen from the North Pole, the reverse is true for every star in the south celestial hemisphere (all of which have a negative declination). Such stars describe circles *below* the horizon and parallel to it so that they never rise above it.

From the North Pole, then, we see only one half the stars in the sky (assuming they are evenly spread over the celestial sphere, which they are, if we consider only those visible to the naked eye). We always see the stars with positive declination and we never see the stars with negative declination.

From the South Pole, the situation is reversed. We always see the stars with negative declination and we never see the stars with positive declination.

Next, imagine yourself back at the North Pole and moving away from it along some particular meridian of longitude toward lower latitudes. Your motion is reflected in the sky, since as you move on the surface of the earth, it seems to you that you remain on top of the sphere with your body vertical and that it is the Celestial Sphere—the entire celestial sphere—that tips.

Suppose you move 10° south of the North Pole. Since the North Pole is at 90° N such a motion brings you to 80° N. At 80° N, the North Celestial Pole seems to have moved 10° away from the zenith and it is now 80° above the northern horizon. In the same way (though you can't see it) the South Celestial Pole has moved 10° away from the nadir and is now 80° below the southern horizon.

This tilt continues as you move toward lower and lower latitudes.

The general rule is that when you are at x° N, the North Celestial Pole is x° above the northern horizon and the South Celestial Pole is x° below the southern horizon. (The

two celestial poles must, of course, always be directly opposite each other on the celestial sphere.)

Again the situation is reversed in the southern hemisphere. As you move away from the South Pole, the South
Celestial Pole tilts downward toward the southern horizon
and the North Celestial Pole tilts upward (unseen) toward
the northern horizon.

The general rule is that when you are at $x°$ S, the South
Celestial Pole is $x°$ above the southern horizon and the
North Celestial Pole is $x°$ below the northern horizon.

At the equator, which is at $0°$, the North Celestial Pole is
$0°$ above the northern horizon and the South Celestial Pole
is $0°$ above the southern horizon. In other words, both celestial poles are exactly at the horizon—at opposite points on
the horizon, of course.

Come back, now, to $80°$ N, where the North Celestial Pole is
$10°$ away from the zenith in the direction of the northern
horizon. The entire celestial sphere is tilted and that includes the celestial equator, half of which is lifted above the
southern horizon and the other half is dropped below the
northern horizon. The maximum height of the celestial
equator is $10°$ above the horizon due south, while the maximum depth is $10°$ below the horizon due north.

Since all the stars make circles parallel to the celestial
equator, all are now making circles that are oblique to the
horizon.

Since the celestial equator dips $10°$ below the northern
horizon at one end of its circle, any star located in the north
celestial hemisphere within $10°$ of the celestial equator—
that is, any star with a declination between $+10°$ and
$0°$—dips below the northern horizon as it moves around the
sky.

On the other hand, since the celestial equator rises $10°$
above the southern horizon at the other end of its circle,
any star located in the south celestial hemisphere within
$10°$ of the celestial equator—that is, any star with a declina-

tion between 0° and −10°—rises above the southern horizon
as it moves around the sky.

Any star with a positive declination of more than +10°
gets closer to the horizon at the northern end of its circle
than at its southern, but never quite sinks below it. Any star
with a negative declination of more than −10° rises closer
to the horizon at the southern end of its circle than at its
northern, but never quite rises above it.

From a stand at 80° N then, we can summarize by saying
that all stars with a positive declination of more than +10°
are always visible in the sky (we're disregarding the occasional presence of the sun, remember) and all stars with a
negative declination of more than −10° are never visible in
the sky. Those stars with a declination between +10° and
−10° are sometimes above the horizon and visible and
sometimes below the horizon and invisible.

We can work this out for any latitude on earth and come
up with a general rule.

If you are standing at $x°$ N on the terrestrial sphere then
all stars with a positive declination of more than $+(90 -
x)°$ are always in the sky, while all stars with a negative
declination of more than $-(90 - x)°$ are never in the sky.
All stars with a declination between $+(90 - x)°$ and
$-(90 - x)°$ rise and set, and so are sometimes in the sky
and sometimes not in the sky.

If you are standing at $x°$ S on the terrestrial sphere, the
situation is symmetrically opposed. All stars with a negative
declination of more than $-(90 - x)°$ are always in the sky.
All stars with a positive declination of more than
$+(90 - x)°$ are never in the sky. All stars with a declination between $-(90 - x)°$ and $+(90 - x)°$ rise and set and
are sometimes in the sky and sometimes not in the sky.

If you are standing on the equator, which is at 0°, then all
stars with a declination between $+(90 - 0)°$ and $-(90 -
0)°$ —that is, between +90° and −90°—rise and set and
are sometimes in the sky and sometimes not in the sky.
But declinations between +90° and −90° are all there is, so

that at the equator, *all* the stars are in the sky at some time or another, all of them making circles that are perpendicular to the horizon.

It is only at the equator that all stars in the sky can be seen at one time or another. (In actual fact, stars near the celestial poles would, as seen from the equator, be always near the horizon and would therefore be difficult to observe —but we are ignoring horizon effects.)

It works the other way around, too. Suppose a star has a declination of +60°. That means it is 30° from the North Celestial Pole. When the North Celestial Pole is more than 30° above the northern horizon, the star must always remain above the horizon. For it to dip below the horizon, it would have to move to a position that is more than 30° from the North Celestial Pole, which is impossible.

The North Celestial Pole is just 30° above the northern horizon when you are standing at 30° N on the surface of the earth. Anywhere on earth from 30° N northward, the star with a declination of +60° is always in the sky. Anywhere on earth from 30° S southward it is never in the sky. Anywhere on earth between 30° N and 30° S, it rises and sets and is sometimes in the sky and sometimes not.

We can present the general rule. If a star has a declination of $+x°$, it is always in the sky from any point on earth north of $(90 - x)°$ N, never in the sky from any point south of $(90 - x)°$ S, and is sometimes in the sky and sometimes not from any point between $(90 - x)°$ N and $(90 - x)°$ S.

If a star has a declination of $-x°$, it is always in the sky from any point of $(90 - x)°$ S, never in the sky from any point north of $(90 - x)°$ N, and is sometimes in the sky and sometimes not from any point between $(90 - x)°$ S and $(90 - x)°$ N.

The corollary to this is that from any point in the northern hemisphere, the North Celestial Pole is always in the sky. It is at 90° N and is therefore always visible from any point north of $(90 - 90)°$ N, or 0°, which is the equator—

while the South Celestial Pole is never in the sky. Contrariwise, from any point in the southern hemisphere, the South Celestial Pole is always in the sky and the North Celestial Pole never is. At the equator, both celestial poles are exactly at the horizon.

Another corollary is that from any point on earth other than the North Pole and the South Pole, any star on the celestial equator is always seen to rise and set and is therefore seen part of the time and not seen the other part.

But now to cases. The declination of Alpha Centauri is −60° 38′, or, since there are 60 minutes of arc to a degree, we can work it out in the decimal system (which I personally prefer) and make the declination −60.63°.

By the rules we have worked out, then, Alpha Centauri is always in the sky for all latitudes south of (90 − 60.63)° S, or 29.37° S. It is *never* in the sky for all latitudes north of 29.37° N. Finally, it rises and sets and is sometimes in the sky and sometimes not in the sky for all latitudes between 29.37° S and 29.37° N.

Next, we must ask ourselves how this relates to the United States.

The line of 29.37° N cuts across Florida at a latitude just north of Daytona Beach. I would estimate, then, that the southern two thirds of Florida offers a view of Alpha Centauri in the sky at certain times.

So far my reviewer is right, but if he undertakes to correct my errors, he is honor-bound to make none of his own. By specifying Florida, he leaves the implication that the "lower" part of the state (to use his geographical term) is the only part of the United States from which Alpha Centauri can be seen.

Not so! The line of 29.37° N also cuts across the southernmost tip of Louisiana, about thirty miles south of New Orleans. From any point in southernmost Louisiana, including the Mississippi Delta region, Alpha Centauri is sometimes visible in the sky.

We're still not through. The line of 29.37° N cuts across the state of Texas at about the latitude of Galveston and San Antonio. From any point in Texas south of those two cities, Alpha Centauri is sometimes visible in the sky.

And we're *still* not through. My reviewer may have forgotten that one of the fifty states is Hawaii and that it is the most southern of all of them. *All* of Hawaii is well south of the 29.37° N mark and therefore Alpha Centauri is in the sky at certain times as viewed from *any* part of the state of Hawaii.

What my reviewer should have said, then, if he had really wanted to be terribly erudite, was that Alpha Centauri was visible from all or part of no fewer than four states of the fifty.

Now, then, how wrong was I?

The area from which Alpha Centauri is visible, at least sometimes, I estimate to be something like this: 36,000 square miles in Florida; 4000 square miles in Louisiana; 40,000 square miles in Texas; and 6400 square miles in Hawaii, for a total of 86,400 square miles. This leaves an area of 3,450,000 square miles in the United States from which Alpha Centauri can never be seen.

In other words, Alpha Centauri can't be seen from 97.6 per cent of the land area of the United States, and it can sometimes be seen from 2.4 per cent. I think, then, that it is a pretty fair approximation to say that Alpha Centauri cannot be seen from the United States in an essay in which I don't want to get into the minutiae of when and where a star can be seen and when and where it cannot be.

In fact, we're not through. A star is seen at a maximum height above the horizon equal to the difference between the latitude at which you are located and the latitude that marks the limit from which the star can be seen.

For instance, the southernmost part of Louisiana is at a latitude of 29.0° N, so that from even the southernmost part of Louisiana, the Mississippi Delta, Alpha Centauri is never more than about 0.4 degrees above the southern horizon.

At even the southernmost point in Texas, the city of Brownsville, Alpha Centauri reaches a maximum height of 3.5° above the southern horizon and at the southernmost point in Florida, Key West, Alpha Centauri is never seen more than 4.3° above the southern horizon.

These are *maximum* heights above the horizon.

Now it is not easy to observe stars that are very close to the horizon. Not only are there often obstructions on the horizon, but even where there are not, there is often a haze.

I should say that, in practical terms, the only portion of the fifty states from which Alpha Centauri is easily visible is Hawaii. Hawaii's area, however, makes up only 0.18 per cent of the nation. Therefore, in a practical sense, Alpha Centauri is not seen in the sky from 99.82 per cent of the United States.

Well, then, what ought I to have said? Ought I to have said, "Alpha Centauri cannot be seen at all from 97.6 per cent of the United States; can be seen in theory and sometimes in fact when one is lucky, in 2.2 per cent of the United States; and can be seen easily in 0.2 per cent of the United States"? Ought I to have said, "Alpha Centauri cannot be seen from the United States, except from the state of Hawaii, and the southern parts of Florida, Louisiana and Texas"?

Or do you think that for the purposes of the article, if I say "Alpha Centauri cannot be seen from the United States," it is worth making a fuss over?

And if my erudite reviewer *is* going to make a fuss over it, how smart is he if he remembers Florida and forgets Hawaii?

THE PLANETS

7.

Just Thirty Years

Some years back there arose the likelihood that my series of six "Lucky Starr" books, teenage novels of science fiction adventure that I wrote originally in the nineteen-fifties under the pseudonym Paul French, would be resurrected, and published in new editions.

"Excellent," said I (for I am never averse to the resurrection of my books), "but the science in them is outdated by now. Therefore, I will insert a short foreword warning readers of this and describing just where the outdating occurs."

The publishers were a bit uneasy about this. They felt it might ruin the sales of the book. I was adamant, however, and I had my way. Soft-cover editions of the books have been published by New American Library and by Fawcett Books, while a hard-cover edition was published by Gregg Press,[1] and in each case my short forewords appeared. The happy ending is that sales did not seem to be in the least impeded.

But how quickly science advances! How quickly statements made in science fiction, in good faith and after careful research, are outmoded, and converted from science fiction into fantasy.

For instance, to get down to cases, exactly thirty years ago this month as I write, *The Magazine of Fantasy and Science Fiction* (where these essays first saw the light of day) was born. How much of what was science fiction at its birth is fantasy now? I don't have the space to review all of science, but suppose we consider one branch that is of particular appeal to science fiction—the planets of the solar system. Suppose we consider them one by one.

Mercury

In 1949, it had been accepted for sixty years that Mercury showed only one face to the sun. Its period of rotation, it was thought, was eighty-eight days—exactly equal to its period of revolution about the sun.

This meant that Mercury had a "sun side" and a "dark side." The sun side was incredibly hot, of course, especially when Mercury was at perihelion and received a hundred times as much solar radiation as earth did. The night side, on the other hand, was in perpetual darkness, and had a temperature little if anything above absolute zero.

In between was a "twilight zone." To be sure, Mercury's orbit was so elliptical that if you stopped to work out the nature of the twilight zone, you would find that almost all

---

[1] My own name was used in these new editions. Paul French has retired.

of it got enough sun at one time and enough darkness at another to end up with an unbearable temperature one way or the other or both.

This was often ignored, though, and Mercury's twilight zone was considered a region of at least bearable temperature—at least in science fiction stories—and human settlements were placed there.

But then microwave astronomy was developed in the decades after World War II and, in the early nineteen-sixties, it was found that microwaves were radiated from the dark side in surprising quantities. The temperature of the dark side had to be well above absolute zero, therefore.

A beam of microwaves could also be sent from earth to Mercury. Striking Mercury, the beam would be reflected and the reflected beam could be picked up back on earth. If the reflecting surface were motionless, the reflected waves would have very much the characteristics of the original beam. If the reflecting surface were moving (as it would be if the planet were rotating) the reflected beam would exhibit changed characteristics, the amount of the change being dependent upon the speed of surface motion.

In 1965, it was discovered from microwave reflections that Mercury rotated not in eighty-eight days after all, but in 58.7 days—just two thirds its period of revolution around the sun.

This meant that the sun side and the dark side of Mercury entered the realm of fantasy. Every part of Mercury experienced both day and night. Each day and each night is eighty-eight earth days long, but there is neither constant day nor constant night anywhere. The surface of Mercury gets hot and it gets cold, but it never gets as hot as the mythical sun side or as cold as the equally mythical dark side.

There went parts of my book *Lucky Starr and the Big Sun of Mercury*.

And what did the surface of Mercury look like? In 1949,

we couldn't say. It was hot, of course, and many were the imagined pools of tin, lead or selenium lying about here and there on the sun side (as in my story *Runaround*).

In 1974, the probe Mariner 10 passed close by Mercury and took photographs that revealed its surface in detail. It looks very much like a larger moon, though it lacks "maria," the wide, relatively flat and unscarred "seas" of the moon. No pools of anything.

Venus

In 1949, we knew virtually nothing about Venus, except for its orbit, its diameter and its brightness. Since it was always obscured by a thick and featureless cloud cover, we didn't know anything about its surface, and we didn't even know its period of rotation.

From its cloud cover, though, which we assumed to be water droplets, we could suppose it was a much wetter and soggier planet than earth. In fact, it even seemed possible that it might have a planetary ocean with little or no dry land. I assumed as much in *Lucky Starr and the Oceans of Venus*.

With a cloud cover and a large ocean, Venus might not be too hot.

In the nineteenth-century theory of solar-system formation called the "nebular hypothesis," it was necessary to suppose that the planets formed from the outside in, so that Mars was older than earth and earth was older than Venus. The nebular hypothesis went by the boards about the turn of the century, but the notion remained in the science fiction mind. It was very common to suppose that Venus was rich in comparatively primitive life. It was still in the dinosaur age, so to speak.

As for rotation, since there was absolutely no way of telling, it was simplest to suppose it rotated in something like twenty-four hours, give or take a little.

By the mid-nineteen-fifties, however, astronomers were beginning to come up with some puzzling observations. Microwaves from Venus seemed to be present in unexpectedly large quantities. Venus might be warmer than expected.

Then, on December 14, 1962, a probe, Mariner 2, flew close by Venus and was able to measure the microwave emission with great precision. It seemed clear that Venus's surface temperature approached an unbelievable 500° C on both the sunlit and the night portion. There couldn't possibly be a drop of liquid water anywhere on the surface of the planet.

Why so hot? The answer lay in the atmosphere. In 1967, a probe, Venyera 4 actually entered Venus's atmosphere, analyzing it as it parachuted down. The atmosphere of Venus, it turned out, was ninety times as dense as that of earth, and 95 per cent of it was carbon dioxide.

Carbon dioxide is transparent to visible light and quite opaque to infrared. Sunlight passes through, is absorbed by the surface and is converted to heat. The hot surface reradiates energy as infrared, which cannot get through the atmosphere. The heat is trapped and Venus's temperature goes up till the infrared is forced through.

Nor are the clouds themselves simply water droplets. There are very likely droplets of sulfuric acid present, too.

What about Venus's period of rotation? Microwaves can penetrate the clouds easily and will be reflected by the surface. Those reflections show Venus to be turning on its axis once in 243.1 earth days and in a retrograde fashion—east to west, rather than earth's west to east. This means that the length of time between sunrise and sunset on Venus is 117 earth days.

Earth

In 1949, the earth was considered to be a rather static place. The land might lift and subside slightly, and shallow arms of the sea might invade and retreat, but the continents stayed put. There had been some theories of "continental drift" but no one of importance believed them.

On the other hand, the ocean floor was beginning to reveal some secrets. The floor was by no means flat and featureless. There was a huge mountain range winding down the Atlantic Ocean and into the other oceans. It was called the Mid-Ocean Ridge.

Making use of sonar soundings, William Maurice Ewing and Bruce Charles Heezen showed in 1953 that, running down the length of the mountain range, was a deep canyon. This was eventually found to exist in all portions of the Mid-Oceanic Ridge, so that it is sometimes called the Great Global Rift.

The rift divided the earth's crust into large "tectonic plates," so called from the Greek word for "carpenter" since they seemed so tightly joined.

In 1960, Harry Hammond Hess presented evidence in favor of "sea-floor spreading." Hot molten rock slowly welled up from great depths into the Great Global Rift in the mid-Atlantic, for instance, and solidified at or near the surface. This upwelling of solidifying rock forced the two plates apart on either side. The plates moved apart at the rate of from 2 to 18 centimeters (1 to 7 inches) a year. As the plates moved apart, South America and Africa moved farther apart.

The continents shift as the tectonic plates move; oceans form; mountain ranges buckle upward; the sea floor buckles downward; volcanoes erupt and earthquakes take place where the plates meet; and so on.

Human beings have invaded the deep. In 1949, a bathyscaphe, a ship capable of maneuvering far below the ocean's surface, had already penetrated 1.4 kilometers (0.85 miles) below the ocean surface.

On January 14, 1960, however, Jacques Piccard and Don Walsh took a bathyscaphe to the bottom of the Marianas

Trench, plumbing 11 kilometers (7 miles) below the ocean surface to the deepest part of the abyss.

Moon

In 1949, as had been true throughout human history, human beings could see only one side of the moon, a side that was airless, waterless, changeless and cratered.

We could dream about the other, hidden side, however. Perhaps, for some reason, it was less forbidding. Even if it weren't, might there not be enough remnants of water and air in the crater shadows or under the moon's surface on either side to support primitive life, at least? Advantage was taken of these notions in occasional science fiction stories.

In 1959, however, a probe, Luna 3, sent back, for the first time ever, photographs of the far side of the moon. Other probes did the same. Eventually, probes orbiting the moon sent back detailed photographs of every part of the moon and the moon could be mapped with almost the detail that the earth could be.

It turned out that the hidden side of the moon was exactly like the visible side: airless, waterless, changeless and cratered. The one difference was that the far side of the moon, like Mercury, lacked the "maria" of the visible side of the moon.

On July 20, 1969, the first human foot was placed on the moon and a few days later the first moon rocks were brought back to earth. Much more has been brought back since and the evidence seems to indicate that not only is there no water on the moon but that there hasn't been any since the early days of the solar system.

In fact, the moon is littered with glassy fragments that seem to indicate that it has been exposed in the past to much greater heat than that to which it is exposed now. Perhaps it had originally had an elliptical orbit that had brought it much closer to the sun at perihelion than it ever gets now and perhaps it had then been captured by earth.

If we ever get samples of Mercury rocks, it will be interesting to compare them with those from the moon.

Mars

In 1949, it was still possible to believe that Mars was covered with an intricate network of canals that could bespeak the presence not only of life, but of intelligent life and of a high, though decadent, civilization. In fact, this became virtually a dogma of science fiction.

To be sure, Mars was smaller than earth, had a far thinner atmosphere and far less water and was far colder, but it had a day that was as long as ours, and an axis that was tipped like ours so that it had seasons like ours—and it had visible ice caps.

The first crack in this picture came on July 14, 1965, when the probe Mariner 4 passed Mars and sent back twenty photographs of the planet.

There were no canals shown. What were shown were craters, rather like those on the moon, and the state of their apparent age seemed to show there could have been little erosion and that there was therefore not much in the way of air or water on Mars.

In 1967, Mariners 6 and 7 passed Mars, and showed that the atmosphere was thinner, dryer and colder than even the most pessimistic preprobe estimates. There could not possibly be any form of advanced life on Mars, let alone intelligent life with great engineering ability. The canals seen by a few astronomers were apparently optical illusions.

In 1971, the probe Mariner 9 went into orbit around Mars and the entire Martian surface was photographed in detail. Though there were no canals, there were enormous volcanoes—one of them, Olympus Mons, far more huge than anything of the sort on earth. Another record was set by Valles Marineris, a canyon that dwarfed earth's Grand Canyon to a toothpick scratch. There were markings, too, that looked precisely like dried river beds.

There was at least some geological life to Mars. Could there be biological life on it, too? Even if only microscopic?

In 1976, the probes Vikings 1 and 2 soft-landed on the Martian surface and tested the soil for signs of microscopic life. The results were rather similar to what might have been expected if life were present but absolutely nothing in the way of organic compounds could be detected.

My stories *David Starr: Space Ranger* and "The Martian Way" were each in part outdated by these discoveries.

Phobos and Deimos

In 1949, the Martian satellites were dim specks of light and nothing more. They were tiny, but that was all we could say.

Some of the later Mars probes took the first close-up photographs of the satellites. They are irregular bodies that look like potatoes, complete with eyes. The longest diameter was 28 kilometers (17 miles) for Phobos, and 16 kilometers (10 miles) for Deimos. Both were thoroughly cratered. Phobos had striations in addition, while Deimos had its craters buried in dust.

The satellites were dark while Mars was reddish. Very likely, Phobos and Deimos are captured asteroids of the kind called "carbonaceous chondrites." These contain considerable quantities of water and organic compounds so that the surfaces of Mars' tiny satellites may prove of greater interest, once they are reached, than the surface of Mars itself.

Asteroids

In 1949, the asteroids were considered to be confined very largely to the asteroid belt and it was a science fiction dogma that the region was littered with debris and was virtually impassable. My first published story, "Marooned off Vesta," dealt with a ship that had been wrecked in the asteroid belt by collision with planetary debris.

To be sure there were occasional exceptions. A few asteroids ("earth-grazers") came in closer than Mars, and in 1948 Icarus had been discovered. It approached the sun more closely than Mercury did. Also, at least one asteroid, Hidalgo, was known to recede as far as the orbit of Saturn.

Over the course of the next thirty years, however, many more asteroids were discovered that penetrate the inner regions of the planetary system. A whole class of "Apollo objects" are now known that approach the sun more closely than Venus does, and in 1978, an asteroid was discovered with an orbit that, at every point, is closer to the sun than earth's orbit is.

In 1977, Charles Kowall, studying photographic plates in search of distant comets, came across an object that seemed to be moving unusually slowly for an asteroid. It turned out to be an object of asteroid size, to be sure, but one with an orbit that, at its closest, was as far from the sun as Saturn was, and, at its farthest, retreated to the distance of Uranus's orbit. He called it Chiron.

It is clear that asteroids are a far more pervasive feature of the solar system than had been thought in 1949. Furthermore, the asteroid belt itself is less dangerous than had been thought. Probes have passed through it without any trouble and without any sign of unusual concentration of matter.

Jupiter

In 1949, it was known that Jupiter was a giant, that it was striped with colors from orange to brown and that it had ammonia and methane impurities in an atmosphere made up largely of hydrogen and helium. Nothing more of its constitution was known than that.

In science fiction stories, it was supposed that under a deep and dense atmosphere there was a solid surface. I took advantage of this belief in my story "Victory Unintentional."

In 1955, active microwave radiation was detected from Jupiter and on December 3, 1973, a probe, Pioneer 10, skimmed its surface. It discovered that Jupiter had a magnetosphere (belts of electrically charged particles outside its atmosphere) that was both far more voluminous and far more densely charged than earth was.

The magnetosphere was deadly, and was large enough to envelope Jupiter's large satellites, which may therefore be unreachable by anything but unmanned probes.

Furthermore, it would appear that the assumption of a sizable solid core must be put aside. Jupiter would seem to be, essentially, a ball of red-hot liquid hydrogen, with a center that may be compressed into solid "metallic hydrogen."

On March 5, 1979, the probe Voyager 1 made a close approach to Jupiter and sent back photographs that showed incredible activity: an atmosphere boiling and twisting in unimaginable storms. One photograph shows what appears to be a thin ring of debris surrounding Jupiter.

Jupiter's Satellites

In 1949, the four large satellites, Io, Europa, Ganymede and Callisto, were known only as dots of light. Their sizes were estimated. Io was moon-sized, Europa a trifle smaller, Ganymede and Callisto considerably larger. Nothing was known of their surfaces, though they were supposed to be smaller versions of Mars. In science fiction, life was frequently placed on their surfaces. I did it in stories such as "The Callistan Menace" and "Christmas on Ganymede."

Once the cratering of Mercury and Mars was discovered, it began to be assumed that the satellites of Jupiter were airless, lifeless, and cratered, too.

The probe, Voyager 1, took the first good closeup pictures of the satellites. Ganymede and Callisto were indeed cratered. The craters were shallow because those satellites were largely icy and the surface didn't have the mechanical strength to support high-walled, deep-centered craters.

The big surprise was that Io and Europa were not cratered.

Europa seemed to be marked by long straight fissures, something like the Martian canals brought to life—except that they are probably cracks in an icy crust. The ice, presumably, fills and blots out any craters that form.

The real surprise was Io. Photos of Io showed there were active volcanoes on it spewing clouds of dust and gas upward. The surface of the satellite must be coated with sulfur lava, which would explain its reddish-yellow color and the haze of sodium around it and through its orbit. It is this lava which has filled in and obliterated any craters that formed.

One small satellite, Amalthea, is inside Io's orbit. It is elongated, with the long axis pointing toward Jupiter, as though tidal effects are pulling it apart. Jupiter's ring is inside Amalthea's orbit.

In 1949, only six small satellites were known to be circling Jupiter beyond Callisto's orbit. Since then the number has risen to eight, possibly nine.

Saturn and Its Satellites

No probes have as yet reached Saturn, so our knowledge of the planet is about what it was in 1949, except that we can suppose that what we have learned about Jupiter is also true of Saturn.[2]

In 1949, the number of satellites known to circle Saturn

---

[2] Since this was written probes have indeed reached Saturn and have shown that the ring system is more complex than we had thought and that its magnetic field is considerably weaker than had been expected.

was nine, as it had been for half a century. In 1967, however, Audouin Dolfuss discovered a tenth satellite, which he named Janus. It circles Saturn more closely than any other satellite, and its orbit lies just outside Saturn's magnificent rings. (I didn't mention Janus, of course, in my book *Lucky Starr and the Rings of Saturn*.)[3]

Uranus

No startling discoveries have been made about Uranus itself since 1949, but in 1977, James L. Elliot and others, who were investigating an occultation of a star by that planet, discovered that the star underwent a pattern of dimming and brightening before Uranus's edge moved in front of it, and the same pattern in reverse after Uranus's opposite edge had passed beyond it.

Apparently, Uranus had rings—thin dark rings not visible to ordinary inspection at that planet's great distance. This, and the even more recent discovery of a ring around Jupiter, now makes it look as though ringed planets may be common, and that every large planet far from its star has them. The remarkable thing about Saturn is not that it has rings but that they are so voluminous and bright.

Neptune

Nothing of significance has been learned about Neptune beyond what was known in 1949.

Pluto

In 1949, Pluto was known only as a dot of light. It was thought to be possibly as large and as massive as earth.

[3] Janus may not exist after all but Saturn-probes have discovered several small satellites near the ring systems.

In 1955, from small but regular brightenings and dimmings, it was found to have a rotation period of 6.4 earth days. The estimate of its size shrank, however, until, in the nineteen-seventies, it was thought to be merely as large and as massive as Mars.

On June 22, 1978, James W. Christy, examining photographs of Pluto, noticed a distinct bump on one side. He examined other photographs and finally decided that Pluto had a satellite, which he called Charon. Pluto and Charon circled each other in 6.4 days, each facing only one side to the other.

From the degree of separation and the time of revolution, it could be calculated that Pluto had a diameter of only 3000 kilometers (1850 miles) and Charon one of 1200 kilometers (750 miles). The two together have only one eighth the mass of our moon.

Summary

Just thirty years has passed since the founding of *The Magazine of Fantasy and Science Fiction* and see what changes have been made in only one small branch of human knowledge.

In those thirty years we have lost the sun side and dark side of Mercury; the oceans of Venus; the canals of Mars; the solid surface of Jupiter; and (possibly) life on any planet in the solar system other than earth.

In those thirty years we have gained the faster rotation of Mercury and the slower rotation of Venus; the hellish heat of Venus; the volcanoes and canyons of Mars; the liquid nature of Jupiter; rings for Jupiter and Uranus; craters for Mercury, Mars, Phobos, Deimos, Ganymede and Callisto; tectonic plates for earth and, possibly, Europa; active volcanoes on Io; additional satellites for Jupiter and Saturn; and a satellite for a shrunken Pluto.

Just thirty years! What will we find out in thirty years more?

Note

The article above was written in March 1979. Since then, time has not stood still, you may be sure.

Saturn's satellites have been studied in detail. Several have been partially mapped. Mimas has a crater that is enormous, considering the size of the satellite. Rhea and Dione are thickly cratered. Enceladus seems to be smooth, but a good look was not obtained by the Voyager probe which passed by in early 1981. Iapetus has one side light and one side dark, but the reason for it is still not known. Titan has an atmosphere much thicker than expected; thicker than earth's is. What's more, the atmosphere is rich in nitrogen.

Then, too, there are Saturn's rings, which have a structure much more complex than anyone dreamed. There may be as many as a thousand subrings making it up, including several in Cassini's division which had been thought to be empty. Some of the subrings are not quite circular and at least one seems to be braided. There are also "spokes" in the rings, dark regions crossing the rings at right angles to their rotation.

A second Voyager probe is on its way.

THE MOON

8.

A Long Day's Journey

Last month my wife, Janet, and I were in upstate New York with a group who were intending to watch the Perseid meteor shower in the small hours of the morning.

Unfortunately, three of the four nights that were devoted to the task were solidly cloud-covered, and on the fourth night the display was not spectacular. Nevertheless, we had a lot of fun, and not the least of it was listening to lectures on astronomical subjects.

One night we were heading out from the hotel to the outlying building where Fred Hess (a marvelous lecturer on matters astronomical) was going to fill us in on ways of predicting eclipses. We were looking forward to it.

In the elevator, an elderly and snappish woman looked at

our outfits with disfavor and said, "You'll freeze to death if you go out like that."

Since I am not particularly sensitive to cold and since I was quite certain that an August evening was not likely to be below 60° F even if the day had been cloudy, I contented myself with smiling benignly. Janet, however, who is more sensitive to cold than I am, looked at her watch uneasily and said, "I don't have time to go back for my sweater."

I was about to assure her that she wouldn't need it, when the harpy cried out, "Are you going to listen to those fairy tales?"

I looked astonished. We Perseid people made up only a small fraction of the total clientele at the resort hotel and there were other activities that had nothing to do with us, but I hadn't heard of any fairy-tale presentations. "Fairy tales?" I asked.

"All that talk about the stars," she said angrily. "Don't listen to it. It's fairy tales."

I'm afraid I laughed, which must have annoyed her, for as we walked off, she decided to escalate the level of insult and applied the very worst epithet she could think of to the innocent lecture on eclipses which we were about to attend. Behind us, her voice rose to a screech. "It's science fiction," she yelled. "*Science fiction!*"

Poor thing! I decided to devote my next essay to something that would sound a lot more science fiction than does the staid and everyday matter of eclipses—not that I imagine for one moment that she would read my essays even if she knew how.

The subject under discussion in this essay is the tidal influence of the moon upon the earth. I have discussed the tides in some detail in an earlier essay ("Time and Tide," in *Asimov on Astronomy*, Doubleday 1974), and I will arbitrarily assume you have read it and remember it.

In that earlier essay, I did spend a few paragraphs on the

way the tides are slowing earth's rotation period, and it is that which I wish to go into in some detail now.

Any of us who played with tops when young[1] know that the rate of their rotation gradually slows and that they eventually wobble, then topple over and are motionless. The rotational energy of a spinning top is gradually depleted and turned into heat through friction of its point with the ground it is spinning on and through the resistance of the air it is turning through. What's more, its tiny store of angular momentum is transferred to the earth's enormous supply.

If the top were spinning without making contact with anything material, and if it were turning in an absolute vacuum, there would be no friction and no way of losing rotational energy or angular momentum. The top would, in that case, spin forever at an undiminishing rate.

If we consider the solid ball of the earth, together with its overlying ocean and atmosphere, as a spinning top, it would seem to represent the ideal case. Earth makes no contact with any material object as it spins and it is surrounded by the vacuum of space.

To be sure, nothing is ever ideal. Interplanetary space isn't quite an absolute vacuum, and the atmosphere and oceans react to the rotation by setting up whirling, energy-consuming currents in air and water. However, so large is earth's supply of rotational energy and angular momentum and so small is the effect of these departures from the ideal that any change in rotation that results from these nonidealities is vanishingly small.

This brings us to the tides. The solid ball of the earth, as it spins, is constantly passing through the two shallow mounds of ocean, one facing the moon, and one facing away from the moon—the tides produced by the fact that different portions of the earth are at slightly different dis-

---

[1] Do tops still exist? I haven't seen anyone playing with a top in years.

tances from the moon and therefore subjected to slightly different intensities of lunar gravitation.

As shorelines pass through the tidal bulges, and as the water moves up the shore and then down, there is a frictional effect. Some of the rotational energy of the earth is converted into heat—and some of its angular momentum vanishes also.[2]

What's more, there are two tidal bulges in the solid earth itself (smaller than the ocean bulges) so that as the earth turns, the rocks heave up a few inches and settle back, heave up a few inches and settle back, over and over, twice a day. Here, too, there is friction and both rotational energy and angular momentum are converted or transferred. Altogether it is estimated that the earth is losing some 20 to 40 billion kilocalories of rotational energy every minute.

As a result of the tidal effect, then, the earth's period of rotation must be constantly slowing; or, to put it in a more mundane way that more immediately impinges upon the consciousness of people, the day must be constantly lengthening.

To be sure, even the colossal loss of 20 to 40 billion kilocalories of rotational energy each minute shrinks to nearly nothing in comparison with the titanic store of rotational energy possessed by the earth. The tidal braking effect is therefore extraordinarily small and it would appear that the day becomes one second longer only after the tides have been exerting their braking effect for 62,500 years. This means that at the end of a century, the day is 0.0016 of a second longer than at the beginning of the century, or, to put it another way, that each day is 0.000000044 of a second longer than the day before.

That's pretty, but can we be sure? Can the lengthening of the day be actually measured?

---

[2] Angular momentum doesn't truly vanish. It can't. It is canceled by an opposite angular momentum, or it is transferred. There is no opposite angular momentum involved here so it must be transferred. But where? We'll take this up in the next chapter.

It can, for we now have atomic clocks that could just about measure such a difference from day to day—and that could certainly do so with ease if we measured the length of several days now and the length of several days next year.

There are complications, though. As clocks grew more accurate, astronomers discovered that the rotation of the earth is not constant and that the earth is, in fact, a rotten timekeeper.

The observed positions, from moment to moment, of bodies such as the moon, the sun, Mercury and Venus, which could be obtained with steadily greater precision as clocks were improved, all showed discrepancies from the theoretical positions they ought to have. What's more, the discrepancies were just about the same for all four bodies. It could not be expected of coincidence that all four bodies would move in unison, so it seemed, instead, that it was the earth's period of rotation that was unsteady.

If the earth's period of rotation slowed slightly, the position of the heavenly bodies would seem to move ahead of theoretical; if the earth's period of rotation speeded slightly, the position of the heavenly bodies would fall behind. Between 1840 and 1920, the rate of earth's rotation slowed by over a second and then it started speeding up again.

Why? Because the earth is not a perfect, unchanging body. There are earthquakes and shiftings of mass within the earth. If the mass shifts, on the average, slightly closer to the center of the earth, the earth's rotation speeds slightly; if the mass shifts slightly farther from the center, the earth's rotation slows slightly.

In fact, as clocks continued to improve, it was found that earth's rotation rate changed with the seasons. In the spring the day is about one twelfth of a second longer than it is in the fall. This is because of shifting mass due to snowfall, seasonal changes in air and water currents and so on.

These changes are all cyclic, however. Seasons and earthquakes will now lengthen, now shorten the day, but in the long run there will be, on the average, no change.

Superimposed on these cyclical changes of a second or more is the much smaller noncyclical change of an increase in the length of the day at a rate of 44 billionths of a second per day. How can one detect that tiny secular change in all that mélange of far larger cyclical changes?

Actually, it isn't difficult.

Suppose that the day has remained constant in length for eons, but has suddenly begun to increase at the rate of a hundredth of a second per year. At the end of a century, the day is one second longer than it had been at the beginning of the century.

Certainly that's not going to make any practical difference in your life, and if all you have is an ordinary watch, you won't even be able to measure the change.

But the differences mount up. Each day in the second year starts 1/100 of a second later than did the equivalent day in the first year, and at the end of the second year, the day starts 365/100, or 3.65 seconds, later than did the day at the beginning of the first year.

Each day of the third year starts 2/100 of a second later than the equivalent day in the first year, so that at the end of the third year the day starts 7.30 + 3.65, or 9.95 seconds later than did the first day of the first year. And so on.

Even though individual days all through the century have been only fractions of a second longer than earlier days, the *cumulative* error from day to day mounts up and by the time a whole century has passed, a particular day would be beginning 2.3 *days* after the moment in time it would have begun had there been no tiny lengthening of the day at all.

Next, suppose that each year, at the same time precisely, something astronomical and noticeable happens—let us say a total eclipse of the sun. Through time immemorial, while the day has been of absolutely constant length, the sun has always been eclipsed, let us say, at 4 P.M. on August 31.

Once the day suddenly begins to lengthen very slowly, the eclipse of the sun begins coming earlier each year by an amount equal to the cumulative error. By the end of the

century, the eclipse would be coming on August 29 at 8:48
A.M.

It doesn't matter what kind of timepiece you have. You
don't need one to tell you that the eclipse is coming earlier;
all you need is a calendar. And from the discrepancy in the
coming of the eclipse, once you eliminate all other possible
causes, you can fairly reason that the day is lengthening at a
rate too small for you to measure directly. In fact, even
without a decent timepiece you can get a good estimate of
the rate.

Of course, an increase of 0.01 of a second per year is large
compared to what really takes place. At the actual rate at
which the earth's day is increasing, the cumulative error in
the course of a hundred years is only thirty-three seconds
and that's not enough to be helpful. This means we must
make use of longer time intervals.

Consider that eclipses of the sun do happen. They don't
happen once a year to the second, but they happen in such
a way that, if we assume the length of a day is constant, we
can calculate backward and decide exactly when an eclipse
ought to have taken place along a certain course on earth's
surface in, say, 585 B.C.

If the length of the day is not constant, then the eclipse
will take place at a different time and the cumulative error
over not one century but twenty-five centuries will be large
enough to detect.

It might be argued that ancient people had only the most
primitive methods for keeping time and that their whole
concept of time recording was different from ours. It would
therefore be risky to deduce anything from what they said
about the time of eclipses.

It is not only time that counts, however. An eclipse of the
sun can be seen only from a small area of the earth, marked
out by a line perhaps 160 kilometers (100 miles) across at
most. If, let us say, an eclipse were to take place only one
hour after the calculated time, the earth would turn in that
interval and at, say, 40° N the eclipse would be seen 1200

kilometers (750 miles) farther west than our calculations would indicate.

Even if we don't completely trust what ancient people may say about the time of an eclipse, we can be sure that they report the *place* of the eclipse and that will tell us what we want to know. From their reports, we know the amount of the cumulative error and, from that, the rate of the lengthening of the day. That is how we know that the earth's day is increasing at the rate of one second every 62,500 years; and is decreasing at that rate, if we imagine time to be going backward, and look into the past.

Determining cumulative errors is one way of measuring the rate of the lengthening of the day. It would be nice to be even more direct, though, and to measure the actual length of an ancient day and show that it was less than twenty-four hours long.

How is that done, though? At a change of 0.0016 of a second per century (increasing as we go into the future, decreasing as we go into the past), it would take a long time to produce a day with a difference in length that would show up on direct measurement.

The day is now exactly twenty-four hours in length, or 86,400 seconds. At the time the Great Pyramid was built some forty-five centuries ago, the day was 86,399.93 seconds long. There is no way we can tell by direct evidence that the pharaohs were living days that were 7/100 of a second shorter than those we are living today.

And as for measuring days in prehistoric times that would seem certainly out of the question.

Yet not so. It can be done. It is not human beings only who keep records, though we are the only ones who do it deliberately.

Corals apparently grow faster in summer than in winter. Their skeletons alternate regions of fast and slow growth and therefore show annual markings we can count. They also grow faster by day than by night and form small daily markings superimposed on the larger annual ones. Naturally, they form some 365 daily ridges a year.

Now let's imagine we are going back in time and studying corals as we go. The length of the year would remain unchanged as we move into the past. (There are factors that would cause it to change, but these are so much smaller than the changes in the length of the day that we make no serious error if we consider the length of the year as constant.) The length of the day grows shorter, however, and there are therefore more of the shorter days in the year. That means the corals ought to show more daily markings superimposed on the annual marking.

Assuming a shortening of the day, of 0.0016 of a second per century as we go back in time, and assuming that rate to be constant, the day should have been 6400 seconds (1.78 hours) shorter 400 million years ago than it is today. The day at that period should therefore have been 22.22 hours long, and there should at that time have been 394.5 such days in the year.

In 1963, the American paleontologist John West Wells, of Cornell University, studied certain fossil corals from the Middle Devonian, fossils that were estimated to be about 400 million years old.

Those fossils showed about 400 daily markings per year, indicating the day to have been 21.9 hours long. Considering the natural uncertainty in the age of the fossils that is pretty good agreement.[3]

Next, let's amuse ourselves by asking another question. The earth reached its present form, more or less, about 4.6 billion years ago. Assuming that, as we go into the past, the day shortens at a constant rate of 0.0016 of a second per century, how long was the day when the earth was first formed?

Under those conditions, the original day was 73,600 sec-

[3] Nevertheless, let's not dismiss the discrepancy. I'll pick that up again in the next chapter.

onds (or 20.4 hours) shorter than it is today. In other
words, the original day, when earth was freshly in existence,
was 3.6 hours long.

Does this sound weirdly impossible?[4] Well, then, let's
compare earth and Jupiter. Jupiter has 318 times the mass
of earth and that mass is, on the average, considerably farther from the axis of rotation, since Jupiter is the larger
body. Both factors contribute to a greater angular momentum of rotation for Jupiter, one that is about 70,000 times
as great as that of earth.

To be sure, Jupiter has four large satellites, two of which
are distinctly more massive than our moon. Each of these
has a tidal effect on Jupiter, which is increased by the fact
that Jupiter's large diameter produces a large drop in gravitational pull across its width.

Doing some quick calculations that take into account the
mass and distance of Jupiter's large satellites, as well as
Jupiter's diameter compared to earth, it seems to me that
the tidal effect of the four satellites on Jupiter is some 1800
times as great as that of the moon on the earth.

And yet, considering Jupiter's enormous angular momentum, it would seem to me that the slowing effect of the satellites on Jupiter's rotation, and the consequent lengthening
of its day, is only one fortieth that of the slowing effect of
the moon on the earth.

Consequently, in the 4.6 billion years since the formation
of the Solar system, Jupiter's day has lengthened by just
about 30 minutes, or 0.5 of an hour. Since Jupiter's day is
now 9.92 hours long, it must have been 9.42 hours long at
the time of formation.

Still, earth's day at the time of formation was only 3.6

---

[4] Well, it is. Since this was written, I received a letter from Charles
Sheffield of Bethesda, Maryland pointing out the simplistic nature
of my calculation. A more sophisticated calculation shows that the
earth's original period of rotation would have been just under thirteen
hours.

hours long, according to my calculations—only two-fifths the length of Jupiter's day at the time of formation. Is that reasonable?

Let's not forget the difference in size between the planets. Jupiter's circumference is 449,000 kilometers (278,600 miles), while earth's is 40,077 kilometers (24,900 miles). If Jupiter turned in 9.42 hours at the beginning, an object on its equator would move at a speed of 13.25 kilometers (8.22 miles) per second. If earth turned in 3.6 hours at the beginning, an object on its equator would move at a speed of 3.1 kilometers (1.9 miles) per second.

As you see, in terms of equatorial speed, the primordial earth would be spinning at less than a quarter of primordial Jupiter's rate. In fact, the primordial earth would be spinning at less than a quarter of Jupiter's rate right now.

Nor would the earth be turning so quickly at the start that it would be in danger of flying apart. Escape velocity from earth is 11.3 kilometers (7.0 miles) per second. Earth would have to turn in about an hour to have its equatorial speed reach the escape velocity.

Earth, then, was born spinning rapidly, and it is owing to the moon's tidal influence that we now have a long day's journey from sunrise to sunrise, one that is nearly seven times the original length.

Suppose that we consider the moon next. Escape velocity from the moon is 2.4 kilometers (1.5 miles) per second. How fast would the moon have to rotate in order that objects at its equator would reach escape velocity and fly away?

The moon's circumference is 10,920 kilometers (6786 miles) and it would have to make a complete revolution in 1.26 hours before it would begin to lose material at the equator. Suppose, just for fun, then, that when it was formed 4.6 billion years ago, it was spinning with a rotation rate of just a trifle over 1.26 hours—just enough for it to hold together.

Suppose, too, that the moon was then located where it is now and that it was subjected to the tidal influence of the earth.

The earth has eighty-one times the mass of the moon so, all things being equal, it would have eighty-one times the tide-producing power. However, the moon is smaller in size than the earth is, and there is a smaller drop in gravitational pull over its lesser width. That tends to negate some of the earth's mass advantage. Even so, the tidal effect of the earth on the moon is 32.5 times that of the moon on the earth.

In addition, the moon's store of angular momentum (if it were rotating in 1.26 hours) would be only one thirty-third of the earth's store right now. Consequently I should judge that the moon would be slowing at a rate some 1000 times that of the earth right now. Its day would be lengthening at the rate of 0.016 of a second per year.

The present sidereal period of rotation of the moon is 27.32 days or 2,360,450 seconds; and if the primordial rotation was 1.26 hours that would be 4536 seconds. To go from the latter to the former at a rate of increase of 0.016 of a second per year (which we will assume will hold constant from year to year) would require about 150 million years, or only one thirtieth of the moon's lifetime.

In other words, as geologic time goes, the moon's rotation period was quickly slowed to its present value.

Why did its rotation not continue to be slowed, until now its period of rotation would be much longer than 27.32 days?

Well, the magic in 27.32 days is that it is precisely equal to the length of time it takes the moon to go around the earth and if the moon both rotates and revolves in that same length of time, it faces one side always to the earth, so that the tidal bulges are frozen in place, with one directly facing the earth and one facing directly away. The moon will then no longer turn through the bulges and there is no longer a tidal slowing effect due to the earth's action upon it. Once it reaches a rotation period equal to its revolution period, it is "locked in" gravitationally and its rotation period no longer changes, except for other, more slowly working, causes.

As you can see, the gravitational effect would work to lock in any small body revolving about some large body, provided the small body isn't too small (the smaller the body the smaller the tidal effect upon it) and provided it isn't too far from the large body (the tidal effect decreases as the cube of increasing distance).

We now know that the two satellites of Mars are gravitationally locked and present only one side to Mars, and we are quite certain this is true also of the five closest satellites to Jupiter.

It used to be thought that Mercury was gravitationally locked to the sun and that it presented one side only to the luminary. However, the sun's tidal effect on Mercury is only about one ninth that of the earth on the moon and, apparently, that is not enough to quite do the job (with Mercury's unusually elliptical orbit increasing the difficulty perhaps). In any case, Mercury rotates in just two thirds of the period of its revolution.

This, too, is a form of gravitational locking and achieves a certain stability. Rotation equal to two thirds the revolution is not as stable as rotation equal to revolution, but the sun's tidal effect is apparently not quite strong enough to knock Mercury out of this lesser level of stability into the greater one.

But now I want to turn to the question of the diminishing store of rotational angular momentum of earth and moon as each slows the other's rotation. That angular momentum must be transferred, but where?

That matter will be taken up in the next chapter.

9.

The Inconstant Moon

When I was twenty years old I was in love for the first time. It was the palest, most feckless and harmless love you can imagine but I was only twenty, and backward for my age.[1]

At any rate, I took the worshipped object to the fairgrounds where there were all sorts of daredevil rides, and paused before the rollercoaster.

I had never been on a rollercoaster, but I knew exactly what it was—in theory. I had heard the high-pitched female screams that rent the air as the vehicle swooped downward, and the manner in which each young woman clung with calculated closeness to the young man next to her had been observed by me.

---

[1] Don't feel bad, Gentle Reader. I've made up for it since.

It occurred to me that if my date and I went on a roller-coaster, she would scream and cling closely to me, and that sensation, I was sure (even though I had not yet experienced it) would be a pleasant one. I therefore suggested the rollercoaster and the young woman, with unruffled composure, agreed.

As we were slowly cranked up to the first peak, I remember speculating on the possibilities of kissing her while she glued herself to me in helpless terror. I even tried to carry out this vile scheme as we topped the rise and started moving downward.

What stopped me was the agonizing discovery that I was possessed of a virulent case (till then unsuspected) of acrophobia, a morbid fear of heights and falling.

It was I who clung to the young lady (who seemed unaffected by either sensation, that of falling or that of being clung to) and I did not enjoy it, either. What I wanted the young lady to be, with every fiber of my being, was the solid earth.

I survived the voyage, but the impression of macho coolness that I had been trying to cultivate was irretrievably ruined and, needless to say, I did not get the girl. (I probably wouldn't have anyway.)

Of course, you mustn't make this out as worse than it is. It is only my own falling that I am averse to and consider to be a bad idea. I don't lose my sleep over other things falling. I have never, for instance, worried about the moon falling.[2]

As it happens, though, the moon is not falling. The fact is, indeed, quite the reverse, which brings me to the topic of this chapter.

In the last chapter I discussed the manner in which the tides sapped the rotational energy of the earth, causing the

---

[2] Not that it's such a bad thing to worry about. Newton did, and, one thing leading to another, he ended with the theory of universal gravitation.

earth's rotation to slow and the day to increase in length at the rate of 1 second every 62,500 years.

I explained that the moon, with a smaller rotational energy than earth, and subjected to a stronger tidal influence from the more massive earth than we are from the less massive moon, had its day lengthen at a more rapid rate. The moon's period of rotation is 27.32 days now, a period that is exactly equal to its period of revolution about the earth (relative to the stars).

With the period of its rotation equal to the period of its revolution, the moon faces one side always to the earth. One tidal bulge on the moon always faces directly toward us and the other directly away from us. The moon does not rotate through the bulges and the tidal action has ceased. Therefore, its day is no longer lengthening after the fashion of the past.

The moon is still subject to a *little* tidal influence from earth, however.

The moon's orbit is slightly elliptical. That means that it is closer to the earth during one half of its orbit than during the other. While the moon is closer than average to the earth, it moves a bit faster than average; while it is farther away, it moves a bit slower.

On the other hand, its rate of rotation is absolutely steady, regardless of the moon's distance from the earth.

While the moon is in the close half of its orbit, its faster orbital speed outpaces its rotational speed and the moon's surface (as viewed from earth) seems to drift very slowly from east to west. In the far half of its orbit, the slower orbital speed falls behind the rotational speed and the moon's surface (again as viewed from earth) seems to drift very slowly from west to east.

This slow oscillation of the moon's surface, first in one direction for two weeks and then in the other direction for two weeks more, is called "libration" from the Latin word for "scales." (The moon seems to be swaying slightly back and forth around an equilibrium point, as scales do when a small weight has been placed in one pan or the other.)

Because of libration, the tidal bulge does move slightly and consumes rotational energy. This tends to damp the libration very slowly and tends to lock the moon more tightly in place. The only way this can happen is for the moon's orbit to become less elliptical and more nearly circular. If the moon's orbit were perfectly circular, the rate of rotation and of revolution would match precisely and libration would end.

The fact that the moon does not revolve in the plane of the earth's equator introduces an off-center pull of the earth's equatorial bulge, which again produces a tidal influence that can be countered by the moon's slowly shifting into the equatorial plane.

These secondary tidal influences I have just described are weaker than the one which gradually slows a world's rotation, so that although there has been plenty of time to slow the moon's rotation to its period of revolution, there has not yet been time to change its orbit into a circular one in the equatorial plane.

Consider Mars' two satellites, though. These were captured, possibly late in Martian history. They would surely have been circling Mars, originally, in rather elliptical and sharply inclined orbits. They are small bodies, though, with very little rotational energy, and Mars' tidal influence has had its way with them. Not only do they face one side eternally toward Mars, but they move in circular orbits in Mars' equatorial plane.

But shouldn't earth's rotation become gravitationally locked, eventually, under the influence of the moon's tidal effects?

We know that the earth's period of rotation is slowing. Because the moon has a smaller tidal effect on earth than earth has on the moon, and because the earth has considerably more rotational energy than the moon ever had, the earth's rate of rotation slows at a much more gradual pace than the moon's did.

Still someday, *someday*, won't earth's rate of rotation slow to the point where it equals the moon's revolution about the earth? Won't one side of the earth always face the moon, just as one side of the moon always faces the earth today? When that happens, the tidal bulge on earth will also be stationary, and neither the earth nor the moon will be subject to the other's tidal influence, and doesn't that mean there will be no further change?

When that happens, the earth might (one would suppose) have a day that was 27.3 days long and earth and moon would circle each other rather like a dumbbell—all in one piece, with the connecting rod being the insubstantial tidal influence.

Well, not quite right. When the dumbbell rotation comes into existence, the period of earth's rotation will not then be 27.32 days long.

To see why not, let's consider.

When rotational energy disappears, it can't *really* disappear, thanks to the law of conservation of energy, but it can (and does) change its form. It becomes heat. The loss of rotational energy is so slow that the heat formed is not significant and just adds, insensibly, to the heat gained from the sun (which must be, and is, radiated away at night).

The earth, as its rotation slows, also loses rotational angular momentum, and this, too, can't really disappear—thanks to the law of conservation of angular momentum. The loss must somehow be made up for by a gain elsewhere.

Angular momentum, without going into the mathematics of it, depends on two things: the average speed of rotation about the axis of all parts of the rotating body, and the average distance from the axis of all parts of the rotating body. The angular momentum goes up or down as the speed increases or decreases, and also goes up or down as the distance increases or decreases.

As the rotational angular momentum goes down through

the loss of rotational speed, thanks to tidal action, this could be made up for, and the law of conservation of angular momentum preserved, if the average distance of all parts of the earth from the axis of rotation were to increase. In other words, all would be well if a slowing earth could expand in size—but it can't. The earth is not going to expand against the pull of its own gravity.

Where does that leave us?

Well, the earth and moon circle each other in a monthly revolution so that there is a revolutionary angular momentum, as well as rotational ones for each body. The two bodies circle the center of gravity of the earth-moon system.

The location of the center of gravity depends on something that we would recognize as the principle of the seesaw. If two people of equal mass were on opposite ends of a seesaw, that seesaw would balance if the fulcrum was under the exact middle of the plank. If one person were more massive than the other, the fulcrum would have to be nearer the more massive person. To be exact, the mass of person A multiplied by his or her distance from the fulcrum must be equal to the mass of person B multiplied by his or her distance from the fulcrum. If person A is ten times as massive as person B, person A must be only one tenth as far from the fulcrum as person B.

Imagine earth and moon at opposite edges of a seesaw and with the fulcrum replaced by "center of gravity." Earth is 81.3 times as massive as the moon. Therefore, the distance from the center of the earth to the center of gravity must be 1/81.3 times as far as the distance from the center of the moon to the center of gravity.

The average distance of the center of the earth from the center of the moon is 484,404 kilometers (238,869 miles). If we take 1/81.3 of that, we get 4728 kilometers (2938 miles).

This means that the center of the earth is 4728 kilometers (2938 miles) from the center of gravity, while the center of the moon is, naturally, 379,676 kilometers (235,931) miles from it. Both moon and earth revolve about this center of gravity once each 27.32 days, the moon making a large circle, and the earth a much smaller one.

In fact, the center of gravity, being only 4728 kilometers (2938 miles) from the earth's center, is closer to the earth's center than the earth's surface is. The center of gravity of the earth-moon system is located 1650 kilometers (1025 miles) *beneath* the surface of the earth.

One can therefore say, without too great a lie, that the moon is revolving about the earth. It is not, however, revolving about the earth's center.

If the moon's orbit were an exact circle, the earth's center would also describe an exact circle though one with only 1/81.3 times the diameter. Actually, the moon's orbit is slightly elliptical, which means that the distance between moon and earth increases and decreases slightly in the course of the month. The position of the center of gravity moves slightly farther and closer to the earth's center in consequence.

At its farthest, the center of gravity of the earth-moon system is 5001 kilometers (3107 miles) from earth's center; and at its closest, it is 4383 kilometers (2724 miles) from the earth's center. Its position, therefore, varies from 1377 to 1995 kilometers (867 to 1240 miles) beneath the surface of the earth.

It is therefore perfectly possible to balance the loss of rotational angular momentum with an equal gain in revolutionary angular momentum. This will take place if the distance of the earth and the moon from the center of gravity increases.

This is another way of saying that as the tidal influence of the moon very gradually slows the earth's rotation, it very gradually increases the moon's distance from us. So, as I said at the beginning of this essay, the moon is not falling, it is rising.

As the moon recedes from us, its apparent angular diameter decreases. In the far past, it was distinctly closer, and, therefore, larger in appearance. In the far future, it will be distinctly farther, and, therefore, smaller in appearance.

That means that in the future, total eclipses of the sun will cease being visible from the surface of the earth. At the present moment, the moon is already somewhat smaller in apparent diameter than the sun is, so that even when the moon is directly in front of the sun, it tends not to cover it all. A thin rim of the sun laps beyond the moon all around and an "annular eclipse" is formed. That's because the average angular diameter of the sun is 0.533° and that of the moon is 0.518°.

If the moon's orbit about the earth were exactly circular and the earth's orbit about the sun were exactly circular, that would be it. There would be only annular eclipses at best and never any total eclipse.

However, earth's orbit is slightly elliptical so that its distance from the sun varies. The sun therefore tends to be a little farther than average during one half the year and a little nearer than average the other half. I've already mentioned that this is true of the moon in its monthly cycle.

The sun is smallest in appearance when it is farthest and its angular diameter is then 0.524°. The moon is largest in appearance when it is nearest and its angular diameter is then 0.558°. There is therefore the possibility of a total eclipse of the sun, when the sun is farther off (and smaller) than usual, or the moon is nearer (and larger) than usual, or both.

As, under tidal influence, the moon recedes, its apparent diameter throughout its orbit will decrease, and, if we assume the sun will remain at its present distance in the meanwhile (as it will) then the time will come when the moon, even at its closest, will have an angular diameter of less than 0.524°. After that, no total eclipse will be visible from the earth's surface at any time.

The moon will have to recede from a closest approach of 356,334 kilometers (221,426 miles) as at present, to a closest approach of 379,455 kilometers (235,793 miles) if it is to appear, even at its largest, no larger than the sun at its smallest. The moon must recede 23,121 kilometers (14,367 miles) for this to happen.

How long will it take the moon to recede by so much?

At the present moment the moon is receding from us at the rate of 3 centimeters (1.2 inches) per year, or roughly 2.5 millimeters (0.1 of an inch) each revolution.

At that rate, it will take the moon about 750 million years to recede that far. Actually, it should take longer, since as the moon recedes its tidal influence weakens and its rate of recession slowly declines. I should suspect it would take closer to a billion years for the recession.

The situation, it appears, would not be so bad. The number of total eclipses per century will slowly decline, the number of annular eclipses will slowly increase, and the duration of the total eclipses that do occur will gradually shorten, but it will be nearly a billion years before the total eclipses cease altogether.

And, for that matter, allowing for stronger tidal influences in the past, it may have been only 600 million years ago, when the first trilobites were evolving, that annular eclipses were impossible. Every time the moon, then slightly larger in appearance than it is now, would pass squarely in front of the sun, the eclipse had to be total.

Let's get back now to the slowing rotation of the earth.

As the earth's rotation rate slows, the moon's distance increases and its time of revolution about the earth also increases. (In addition, tidal influences will see to it that the moon's period of rotation will slow in time with the slowing of its period of revolution.)

Thus, by the time that the moon has receded to a distance which will make total eclipses impossible, the month will no longer be 27.32 days long relative to the stars, but will be 29.98 days long. And as the moon continues to recede, the month will continue to grow longer.

By the time the earth's period of rotation has lengthened to 27.32 days—the length of the present period of revolution of the moon—the then period of revolution will be substantially longer, and the earth's rotation will have to continue to slow before the dumbbell rotation will be set up.

Is it possible the earth will never catch up? That no matter how slowly it rotates, the moon will retreat so far that its period of revolution will always be longer?

No, the earth's rotation *will* catch up. When the earth's rotation has slowed to the point where the day is equal to 47 present days, the moon will have receded so far that its period of revolution will also be equal to 47 present days.

At that time, the distance of the moon from the earth will be, on the average, 551,620 kilometers (342,780 miles) and its apparent angular diameter will be about 0.361°.

Then we will have earth and moon revolving about each other dumbbell fashion, and if there were no outside interference, that would continue forever.

But there *is* outside interference. There is the sun.

The sun exerts a tidal effect on the earth, as the moon does, but to a different extent. The tidal effect on the earth by each of two bodies varies directly with the mass of the two bodies and inversely as the *cube* of their distances from the earth.

The sun's mass is 27 million times that of the moon. However, the sun's distance from the earth is 389.17 times the moon's distance from the earth and the cube of 389.17 is about 58,950,000. If we divide 27,000,000 by 58,950,000, we find that the sun's tidal effect on the earth is only about 0.46 that of the moon.

The tidal effect on earth of all bodies other than the sun and the moon is insignificant. We can say, then, that the total tidal effect on earth is roughly two thirds moon-caused and one third sun-caused.

The lengthening of the day by one second in 62,500 years is the result of the tidal effect of moon and sun combined, and it is the combined effect that is balanced by the recession of the moon.

Once the earth and moon reach their dumbbell revolution, however, the moon's tidal effect virtually vanishes. That leaves the sun's tidal effect alone in the field. Without going into details, the sun's tidal influence on earth and moon together is such as to speed the rotation of both bodies and balance that increase in rotational angular momentum by a decrease in revolutionary angular momentum.

In other words, the moon will begin to spiral closer to the earth. (*Then*, finally, it will be falling.) The moon will come closer and closer to the earth and there is apparently no limit on how close it can come—except that it will never actually crash into the earth.

As the moon approaches the earth, the tidal effect of the earth on the moon will increase. By the time the center of the moon approaches to only about 15,000 kilometers (9600 miles) from the center of the earth, and the surface of the moon is only 7400 kilometers (4600 miles) from the surface of the earth, the moon will be revolving about the earth once every 5.3 hours. By then, the earth's tidal effect on the nearby moon will be fifteen thousand times as great as it is now, or five hundred thousand times the intensity of the moon's present tidal effect upon us.

Under those conditions, earth's tidal influence will begin to pull the moon apart into a number of sizable fractions. These will collide and fragment and gradually, through continuing tidal effects, spread out over the entire orbit of the moon, forming a flat, circular ring in the equatorial plane of the earth.

In short, the earth will acquire a ring, smaller in actual extent than Saturn's but much denser in material, and much brighter, since earth's rings will be much closer to the sun (despite the fact that the moon rings will be made up of dark rock rather than the ice of Saturn's rings).

Will there be human beings present on earth to watch those beautiful rings? Not unless we have long since left earth and are watching from a distance.

The moon's tidal effect upon earth at the time of its own

breakup would be fifteen thousand times what it is now. That would not be enough to break up the earth, since it would be far less than earth's tidal effect on the moon, and since the earth would be held together by a stronger gravitational pull.

The moon's tidal effect would, however, be strong enough to create tides several kilometers high and would send the oceans washing over the continents from one end to the other.

After the moon's breakup, the tidal effect on us, coming, as it would, from all directions, would cancel out and disappear, to be sure, but by then, after millions of years of enormous tides, the damage would have been done. It's hard to see how land life, or perhaps any life, could survive under such conditions.

That point is, however, academic, since the earth would have ceased to be habitable long before the moon started approaching again.

Let's go back to the dumbbell rotation, with earth's day 47 present days in length.

Imagine what it would mean having the sun shine down for some 560 hours between rising and setting. It shines for longer than that at one time in the regions of the pole, of course, but the sun is then skimming the horizon. Imagine 560 hours between sunrise and sunset in the tropics with the sun riding high in the sky. There's no doubt that by midafternoon the oceans would be nearly (if not quite) boiling.

That alone would put into serious question the habitability of the earth, without our having to regard the Antarctic conditions to which the earth would sink in the course of a 560-hour-long night.

The alternation in temperature between prolonged day and prolonged night would make it very difficult, if not impossible, for life to maintain a foothold on the planet.

And yet that point is academic, too, as we will find when we calculate the time it would take for the moon to recede to a distance at which its period of revolution would be forty-seven days. It will by that time have receded 167,200 kilometers (104,000 miles) beyond its present distance.

If its present rate of recession of three centimeters a year were to continue year after year, then it would take something like 55.7 billion years for the moon to recede to the point where earth and moon were revolving dumbbell fashion.

· The recession will not continue at its present rate, however. As the moon recedes, its tidal effect on earth diminishes earth's rate of rotational slowdown decreases and the moon's rate of recession would decrease, too.

My guess is that it would take at least 70 billion years for the dumbbell situation to be achieved.

And of what significance is such a period of time, when in 7 billion years (just one tenth the time required for reaching the dumbbell situation) the sun will expand into a red giant and both earth and moon will be physically destroyed?

In the course of the 7 billion years before the earth is made uninhabitable by the heating and expanding sun, the earth's period of rotation will have slowed down only to the point where the day would be fifty-five hours long. In fact, allowing for the slow decrease in intensity of the moon's tidal effect, I suspect the day would be forty-eight hours long, or just twice its present length.

It will then get hotter during the day and colder during the night than it does now, and earth won't be as pleasant a place then as it is today; but it will still be habitable, if that were all we had to worry about.

But there *is* the sun, and assuming that humanity survives for 7 billion years, it will be the expanding sun that will drive us away from our planet and not the slowing rotation.

THE
ELEMENTS

10.
The Useless Metal

When the Three Mile Island nuclear power plant went wrong, I came to certain conclusions and found, as often happens, that I was out of step with the world.

The predominant sentiment seemed to be: "Aha! Scientists told us it couldn't happen, but it *did*. So much for those smarty-pants scientists. Now let's tear down the nuclear age."

And yet that's not what really happened. Scientists never said things couldn't go wrong. They said enough safety measures had been taken to make the chance of real damage extraordinarily small.

What the antinuclear people said was something like this: "Wait! An accident will take place and hundreds of thousands of people will be killed outright and millions will get cancer and thousands of square miles of land will be forever useless."

So? Three Mile Island seems to have been poorly designed to begin with. People in charge seem to have disregarded certain warning signals and to have been unnecessarily careless. There were mechanical failures followed by human error. There was even theoretical insufficiency since a hydrogen bubble formed that no one had ever predicted.

In other words, it was practically a worst-possible-case kind of accident. What were the consequences?

The power station was put out of action and will stay out of action for a long, long time, but not one person was killed and there is no clear evidence that anyone was hurt, for radiation escape was low. There may be an additional case of cancer or two as a result and, while I don't want to minimize this, the number of cancer cases will be far less than will be caused in the same area by tobacco smoking and automobile exhaust.

It seemed to me, then, that the Three Mile Island incident was a case where the scientists' predictions proved correct and those of the antinuclear people incorrect. And yet the incident was instantly labelled a "catastrophe" by the media and the antinuclears. What would they have called it, I wonder, if one person had been killed?

In any case, when the Philadelphia *Inquirer* asked me to write a piece stating my views on the matter, I wrote a sardonic article for the April 15, 1979, issue. My pronuclear views ran side by side with an antinuclear article by George Wald.

Two weeks later, I was in Philadelphia and a young woman stopped me and said, rather sadly, "I was *sure* that you of all people would be on the antinuclear side. You're so liberal."

That saddened me. I am certainly a liberal, but that doesn't mean I automatically plug in to the official liberal viewpoint. I like to think for myself—a prejudice of mine of long standing.

Still, all that brooding on the subject reminded me at last that I have never written an *F & SF* essay on uranium. So here goes:

To begin at the beginning, there is a mineral called blende, from a German word meaning "to blind" or "to deceive." (Many mineralogical terms are German because Germany led the world in metallurgy in the Middle Ages.)

The reason for the use of the word is that blende looks like galena, a lead ore, but it yields no lead and therefore it deceives miners.

Actually, blende is mostly zinc sulfide and it has become an important zinc ore. It is now more commonly called sphalerite, from a Greek word meaning "treacherous," which still harps on its deceitful nature.

There are other varieties of blende, differing among themselves in appearance, in one way or another. One is called pitchblende, not because it is in any way pitchy or tarry, but only because it is a glossy black in color; black as pitch, in other words.

Pitchblende is met up with in conjunction with silver, lead and copper ores in Germany and Czechoslovakia. The early mineralogists considered it an ore of zinc and iron.

One place where pitchblende occurs is at the silver mines in St. Joachimsthal (St. Joachim's Valley) in Czechoslovakia, 120 kilometers (70 miles) west of Prague, just at the East German border. (The place is now called Jachymov by the Czechs.)

The spot is of particular interest to Americans because about the year 1500, coins were struck that were made out of the silver from the mines there and that were therefore called Joachimsthalers, or Thalers for short. Other coins similar in size and value were also called that and eventually the name was used, in 1794, by the infant United States for its unit of currency—which we call "dollars." (St. Joachim, if you want to know, was, according to legend, the father of the Virgin Mary.)

One person who interested himself in pitchblende was the German chemist Martin Heinrich Klaproth (1743–1817). In 1789, he obtained a yellow substance from pitchblende which he rightly decided was an oxide of a new metal.

At that time, the tradition of associating the metals and the planets was still strong. In one case, the metal quicksilver was so closely associated with the planet Mercury that it actually received the planetary name as its own, at least in English.

As it happened, eight years earlier, the German-British astronomer William Herschel (1738–1822) had discovered a new planet and had named it Uranus, after Ouranos, the god of the sky in the Greek myths, and the father of Kronos (Saturn). Klaproth decided to name the new metal after the new planet and he named it uranium.

As it turned out, pitchblende is largely a mixture of uranium oxides and it is now called uraninite.

Klaproth then tried to react the yellow uranium oxide (actually uranium trioxide, UO_3) with charcoal. The carbon atoms of the charcoal, he expected, would combine with the oxygen in the uranium trioxide, leaving behind metallic uranium. He did obtain a black powder with a metallic luster and assumed that was uranium metal. So did everyone else at the time. Actually, the carbon had combined with only one oxygen atom from each molecule, leaving behind the blackish uranium dioxide, UO_2.

In 1841, a French chemist, Eugène Peligot (1811–90), realized there was something odd about the "uranium metal." When he conducted certain chemical reactions, the uranium at the beginning and at the end didn't add up correctly. Apparently, he was counting in some nonuranium atoms as uranium. He grew suspicious that what he considered uranium metal was really an oxide and contained oxygen atoms in addition to uranium.

He therefore decided to prepare uranium metal by a different procedure. He started with uranium tetrachloride (UCl_4) and tried to tear the chlorine atoms away by using something a good deal more active than charcoal. He used metallic potassium, not at all a comfortable substance to deal with, but the cautious Peligot performed the experiment carefully enough to suffer no harm.

The chlorine atoms were successfully removed, all of them, and left behind was a black powder with properties quite different from those of Klaproth's black powder. This time, the powder was the metal itself. Peligot was the first to isolate uranium—a half century after it had been *thought* to have been isolated.

No one cared much about this, however, except a few chemists. Uranium was a thoroughly useless metal and no one, except for those same few chemists, ever thought of it— or even heard of it.

In the early nineteenth century, it came to be accepted that the various elements were made up of atoms, and that those atoms had characteristic differences in mass. By following the events in various chemical reactions it was possible to judge the relative masses of the different kinds of atoms ("atomic weights"), but it was also possible to make mistakes.

Counting the mass of the hydrogen atom (the lightest one) as 1, the atomic weight of uranium was taken to be about 116 around the middle of the nineteenth century.

This meant that uranium atoms were fairly massive, but by no means unusually so. Uranium atoms, it was thought, were a little more massive than silver atoms and a little less massive than tin atoms.

The most massive atoms were, at that time, thought to be those of bismuth, the atomic weight of which was 209. The bismuth atom, in other words, was thought to be 1.8 times as massive as the uranium atom.

In 1869, however, the Russian chemist Dmitri Ivanovich Mendeléev (1834–1917) was working out the periodic table. He was arranging the elements in the order of their atomic weight and in a system of rows and columns that divided them into natural families, with all members of a given family showing similar properties.

In some cases, Mendeléev came across an element that didn't fit this neat family arrangement. Rather than assume his whole notion was wrong, he wondered if the atomic weights might in those cases be mistaken. For instance, uranium's properties didn't fit if it were pushed into the atomic-weight-116 slot. If its atomic weight were doubled, it did fit.

Starting from this new slant, it was easy to reinterpret the experimental findings and show that it actually did make more sense to suppose the atomic weight of uranium to be in the neighborhood of 240 (the best current figure is 238.03).

This was about 1871, and for the first time the useless metal uranium gained an interesting distinction. It had a higher atomic weight than any other known element. Its atoms were 1.14 times as massive as those of bismuth.

For over a century now, it has retained that distinction, in a way. To be sure, atoms more massive than those of uranium have been dealt with, but they were all formed in the laboratory and they don't survive for long—certainly not for geological periods.

We can put it this way. Of all the atoms present in the earth's crust at the time of its formation, the most massive that are still to be found in the earth's crust today in more than vanishing traces are those of uranium. What's more, they are the most massive that *can* exist (though, of course, this was not understood in 1871).

The position of uranium at the end of the list of elements was interesting—to chemists. To the world generally, it remained a useless metal and of no account.

So things stood till 1896.

The year before that, Wilhelm Konrad Roentgen (1845–

1923) had discovered X rays and had suddenly become world-famous. X rays became the hottest thing in science and every scientist wanted to investigate the new phenomenon.

Roentgen's X rays had issued from a cathode-ray tube, and the cathode rays (streams of speeding electrons, it was soon discovered) produced fluorescent spots on the glass, and it was from those spots that the X rays were given off. Furthermore, the X rays were detected by the fact that they induced fluorescence in certain chemicals. Therefore, there might be some connection between X rays and fluorescence generally.

(Fluorescence, by the way, takes place when atoms are excited in some way and are raised to a higher energy level. When the atoms fall back to normal the energy is given off as visible light. Sometimes the fall to normal takes time and visible light is given off even when the exciting phenomenon is removed. The light is then called phosphorescence.)

As it happened, a French physicist, Antoine Henri Becquerel (1852–1908), was particularly interested in fluorescent substances, as his father had been before him. It occurred to him that fluorescent substances might be emitting X rays along with visible light. It seemed to him to be worth checking the matter.

To do that, he planned to make use of photographic plates, well wrapped in black coverings. Light could not get through the coverings and even exposure to sunlight would not succeed in fogging the plates. He would put the fluorescent substance on the covered plate and if the fluorescence was ordinary light only, the plate still would not be fogged. If, however, the fluorescence contained X rays, which had the property of passing through a reasonable thickness of matter, they would pass through the covering and fog the photographic plate.

Becquerel tried this on a number of different fluorescent substances with negative results; that is, the photographic plates remained resolutely unfogged. One fluorescent sub-

stance, in which Becquerel's father had been particularly interested, was potassium uranyl sulfate, a substance made up
of complex molecules that contained one uranium atom in
each molecule.

That alone, of the fluorescent substances Becquerel tried,
seemed to give a positive result. After some exposure to the
sun, the photographic plate, on development, showed some
fogging. Becquerel's heart beat faster and his hopes
climbed. He hadn't had a chance to do much exposing because it had been a largely cloudy day, but as soon as the
weather cleared, he planned to do a better job, give it a
good slug of exposure and check the matter beyond all
doubt.

Of course you know what had to happen. Paris settled
down for a long siege of wet weather and there was no sunlight. Becquerel had obtained new photographic plates, well
wrapped, and he had no chance to use them. So he put
them in the drawer, put the potassium uranyl sulfate in the
drawer with them and waited for the sun.

As the days passed and the clouds persisted, Becquerel
got so upset that he decided he had to do something. He
might as well develop the new plates and see if he had had
some lingering phosphorescence that included X rays. He
developed the plates and was stupefied. They were tremen-
dously fogged, almost as though he had exposed them un-
covered, to sunlight.

Whatever was coming out of the potassium uranyl sulfate, it could pass through black paper and it didn't require
prior excitation by the sun. In fact, it didn't require fluorescence, for samples of potassium uranyl sulfate that had been
on reagent shelves away from sunlight for indefinite periods
also fogged the plates. What's more, uranium compounds
that were *not* fluorescent at all also fogged the plates. What
was still more, the amount of fogging depended on the
amount of uranium present and not on that of any of the
other atoms.

It was uranium, and uranium specifically, that gave rise to these X-raylike radiations.

Almost at once, a brilliant Polish-French chemist, Marie Sklodowska Curie (1867–1934), began to study the phenomenon, and she termed it "radioactivity." Uranium, in other words, was radioactive. Curie discovered that another element, thorium, with atoms nearly as massive as those of uranium (the atomic weight of thorium is 232) was also radioactive.

The fact of radioactivity was glamorous. Nothing like that had ever been detected before.

The implications were even more important than the fact itself.

Radioactive atoms were giving off some radiations that were like X rays but were even more penetrating. These were "gamma rays."

But radioactive atoms were also giving off something else, streams of particles that were much smaller than any atoms. This was the final proof of something that was just coming to be suspected: that atoms were not the ultimate particles of matter they had been taken to be since they were first proposed in 1803 (and, in fact, since they had first been conceived by the ancient Greeks twenty-two centuries earlier). Atoms were made up of still smaller "subatomic particles."

When a uranium atom or a thorium atom gave off a subatomic particle, that changed its structure and made of it an atom of a new element. It was, after all, possible to transmute one element into another, as the old alchemists had thought, but under far different conditions than any that the alchemists could have imagined.

Uranium and thorium, as it happened, changed spontaneously into lead. (The alchemists had tried to transmute lead into gold and here was the new transmutation doing the job of *forming* lead, for goodness' sake.)

The change took place, however, very slowly. Half of all the uranium in existence on earth (or half of any portion of it which one was dealing with) was converted to lead only after 4.5 billion years. Half of what was left would be converted to lead only after another 4.5 billion years, and so on. To express this, we say that the half-life of uranium is 4.5 billion years.

In the case of thorium, the half-life is 14 billion years.

Consequently, of the uranium or thorium that had existed on earth when the planet was just formed, half of the uranium and four fifths of the thorium is still in existence today.

Through researches involving radioactivity, the New Zealand-born physicist Ernest Rutherford (1871–1937) was able to show, in 1906, that an atom consisted of a tiny, massive nucleus at the core, surrounded by one or more relatively light electrons. The nucleus carried a positive electric charge, and the electrons negative electric charges. The charges balanced so that the atom, as a whole, was electrically neutral.

In 1913, the English physicist Henry Gwyn-Jeffreys Moseley (1887–1915) showed from the radiation produced by bombarding various metals with X rays that every element had a characteristic positive electric charge on its nucleus. These were whole-number multiples of the charge on the hydrogen nucleus and it was called the "atomic number."

It turned out that thorium had a high atomic number of 90 and uranium a still higher one of 92. Uranium had, in fact, a higher atomic number than any other element found naturally on earth.

This seemed to make sense. Positive electric charges repelled each other and if enough were piled onto a single nucleus it might cause the nucleus to break apart through the mutual repulsion of the charges. Thorium and uranium could just barely hold their nuclei together and broke down slowly. Any element with an atomic number greater than 92 would break down more quickly. If any had been present when the earth was formed, all would be gone by now.

In fact, a number of elements with atomic numbers less than 90 and 92 might be too short-lived to exist. At the time the radioactivity of uranium was discovered, the element of the highest atomic weight known, other than uranium and thorium, was bismuth, and, as it eventually turned out, it had an atomic number of 83.

Could it be that no element with an atomic number greater than 83 was stable, and that of those beyond-bismuth elements, only thorium and uranium were sufficiently close to stability to last through all of earth's history so far?

The answer to that is yes.

It might be thought, then, that uranium, starting with 1896, would be *the* glamour element of the list, sharing the spotlight just a bit with thorium. After all, it was radioactive, it underwent spontaneous transmutation, it was the most complex of all the atoms and had a record high atomic number. What more could one ask?

And yet uranium, after a very short stay in the limelight, sank downward into comparative oblivion once again.

It happened this way . . .

Since uranium had such a long half-life, very few atoms were breaking down at any one particular moment and the amount of radioactivity it produced should be very low. Yet when uranium minerals were tested for radioactivity, it was found that the radioactivity detected was far higher than could be accounted for by the uranium present.

What's more, if uranium compounds were separated from the minerals and refined to a high degree of purity, the radioactivity of those uranium compounds was found to be low —about as low as it ought to be.

That meant that present in the uranium minerals were substances that were more radioactive than uranium—much more radioactive. But how could that be? If those sub-

stances were that radioactive, they should have broken
down long ago and be all gone. What were they doing in
the minerals?

What's more, it turned out that the pure uranium compound, freshly isolated from the minerals and hardly radioactive at all, grew steadily more radioactive as it stood.

What was happening was that the uranium (atomic number 92) wasn't being converted to lead (atomic number 82)
in one fell swoop. Instead, the uranium was converted
through a series of steps to lead, by way of a whole series of
elements of intermediate atomic number. It was these intermediate elements that were more radioactive than uranium
and they *would* break down and vanish if fresh supplies
weren't formed continually from the further breakdown of
uranium.

Of course, if an element is formed very slowly and breaks
down very rapidly, there is very little of it present at any
one time. Under ordinary circumstances there would be far
too little of it present to be detectable or isolable.

The circumstances are not ordinary. The intermediate elements are giving off radiation that makes it possible to detect even infrasmall quantities.

Curie and her husband, Pierre (1859–1906), set about
isolating some of these radioactive intermediates. They subjected pitchblende to chemical reactions that would separate the different elements present, and always followed the
trail of the radioactivity. Whenever the reaction succeeded
in producing a solution or a precipitate in which the radioactive radiation seemed to be concentrated, they worked on
that solution or that precipitate.

Step by step, they worked their material down to smaller
and smaller quantities of more and more richly radioactive
material. In July 1898 they isolated a few pinches of powder
containing a new element hundreds of times as radioactive
as uranium. This they called polonium after Curie's native
land, and its atomic number is 84.

Working on, they detected, in December 1898, a still

more radioactive substance with an atomic number that eventually proved to be 88. They named it radium because of the overwhelming strength of its radioactivity. Its half-life is 1622 years and it is 3 million times as radioactive as uranium and 8.7 million times as radioactive as thorium.

The Curies had so small a quantity of radium to begin with that they could detect its presence only by the radiations. That was enough, in theory, but they wanted an actual quantity that they could weigh and show in the time-honored way of establishing the existence of a new element.

For that they had to start with tons of waste slag from the mines at St. Joachimsthal. The mine owners were delighted to let the crazy chemists have all they wanted, provided those chemists paid the shipping costs. The Curies got eight tons.

By 1902, they had succeeded in producing a tenth of a gram of radium after several thousand steps of purification, and eventually they obtained a full gram.

Radium stole the show. For forty years, when one mentioned radioactivity, one thought of radium. It was the wonder substance *par excellence*, and people or institutions who could gain a tiny quantity to experiment with felt themselves fortunate indeed.

As for uranium, it instantly dropped back out of the limelight once more. It was only the dull parent substance, interesting (if at all) only for the sake of its glamorous daughter.

And yet who hears of radium today? Who cares about it? It is utterly uninteresting and it is uranium that is the wonder of the world.

The ugly duckling had become a vulture.

I'll explain in the next chapter.

11.
Neutrality!

The science fiction writer Lester del Rey is, like myself, a member of a small group called the Trap-Door Spiders. Once a month we attend a dinner, and the usual routine is that I get a taxi near my place, direct the driver to Lester's place, pick him up and then go to the dinner.

Usually, Lester is waiting in front of the door of his apartment house. This time, however, I was a little early and he had not yet come down.

That didn't bother me. I just called to the doorman, "Sir, please ring Lester del Rey and tell him his taxi awaits."

At this, the taxi driver, who till then had confined himself to an occasional wheeze, sat up excitedly and cried out, "Lester del Rey? You know Lester del Rey?"

"He's a friend of mine," I said, with quiet pride.

"I listen to him all the time on the late-night shows!" said the driver, in clear awe. (Lester has been a frequent guest on such shows since there are few people who can sound so authoritative on so vast a number of subjects, and still fewer who hesitate less to do so.)

"Well, there he is," I said.

As Lester approached the cab, the driver said to me gruffly, "Move to the other side of the seat, you. I want to be able to talk to Mr. del Rey."

I moved. Lester took his seat. The driver fawned all over him and Lester accepted it with a visible expansion of his cephalic diameter. They talked briskly the entire trip and Lester did not bother to introduce me.

Nor did I try to introduce myself. This was not out of any sudden attack of diffidence or modesty, you understand. It was just that, being a morning person, I am never on late-night shows, and so I was quite certain that the driver had never heard of me. I didn't want to contribute further to Lester's cranial swelling by demonstrating that fact.

Besides, thought I, it's not always the glamour of the moment that counts. Look at radium.

In the last chapter, you will remember, I had reached the point where radium had become a superstar among elements, with uranium all but ignored except as its dull progenitor. But, of course, conditions didn't remain so.

The discovery of radioactivity and of the streams of subatomic particles given off by radioactive elements had led to an understanding of the structure of the atom.

Through the work of the New Zealand-born Ernest Rutherford (1871–1937), it became clear by 1911 that almost all the mass of the atom was concentrated in a nucleus at the center. The nucleus was only 1/100,000 the diameter of the atom itself. What made up the vast bulk of the atom was a cloud of low-mass electrons.

The nature of the atom could be altered if the nucleus were banged about with sufficient energy to alter *its* structure. This was not likely under ordinary circumstances, however. At everyday temperatures, atoms striking each other do so with energy far, far less than is required to break through the electronic barriers and to allow one nucleus to strike another.

Radioactive atoms, however, give off subatomic particles no larger than electrons or nuclei in size, and these could slip through the electron barrier and into the depths of the atom. This is especially true of the "alpha particles," which are as massive as helium atoms (indeed, it eventually turned out that they were naked helium nuclei). If the alpha particle just happened to be aimed correctly, it would penetrate an atom and strike its nucleus. In doing so, it might rearrange the nuclear structure and change its identity. This would be a "nuclear reaction."

The first deliberate nuclear reaction induced in this way came in 1919. It was carried through by Rutherford, who managed to transform nitrogen atoms into oxygen atoms.

Rutherford proceeded to bombard atoms of many different varieties with alpha particles in order to induce further nuclear reactions and, in the process, learn more about nuclear structure and the fundamental properties of matter.

There was a catch, though. Alpha particles were charged with positive electricity and so were atomic nuclei. Similar electric charges repel each other so that, as an alpha particle approached a nucleus, the particle was repelled, lost velocity and energy and became less capable of inducing a nuclear reaction.

The more massive the atomic nucleus, the greater its positive charge and the greater its repelling effect. For nuclei more massive than that of potassium (with a nucleus carrying a charge of +19) no alpha particle found in nature possessed enough energy even to strike the nucleus, let alone rearrange it.

One alternative was to use protons as subatomic missiles. Since protons are hydrogen nuclei, they are easy to obtain. They have an electric charge of +1, only half that of the alpha particle, so that the protons are repelled less intensely and, all things being equal, can more easily strike a nucleus.

All things are not equal, however. A proton has only one fourth the mass of an alpha particle and can disturb the nucleus correspondingly less.

But then, beginning in 1929, devices were developed that accelerated charged particles, particularly protons, and imparted to them far more energy than was found naturally in connection with radioactive atoms. The most successful device of this sort was the cyclotron, invented by the American physicist Ernest Orlando Lawrence (1901–58) in 1931. After that, the art of bringing about nuclear reactions by bombardment with subatomic particles went into high gear.

It was clear that nuclear reactions produced far more energy per mass of reacting materials than chemical reactions did. (Chemical reactions involve only the outer electron cloud of the atoms.) It didn't seem likely, however, that such nuclear energy could be tapped by human beings. Unfortunately, atomic nuclei are so incredibly tiny and make up so minute a portion of the atomic volume that most subatomic particles, fired at random (as they had to be), missed the nuclei. This meant that the energy expended on accelerating the particles was far greater than the nuclear energy produced by the vanishingly small percentage of those particles that scored direct hits on the nuclei.

But science doesn't stand still. In 1930, evidence was obtained to the effect that when beryllium atoms were exposed to alpha rays, something—call it N—emerged which could induce nuclear reactions. It was just as though N were a stream of subatomic particles.

The trouble was, though, that all the devices that served to detect subatomic particles detected nothing at all in the case of N.

This might not be a mystery. What such devices detected whenever they reacted to the presence of subatomic particles was not the particles themselves but the electric charges on the particles.

In 1932, the English physicist James Chadwick (1891–1974) pointed out that N could be explained easily if one were to suppose that it consisted of a stream of particles that were as massive as protons but that lacked any electric charge at all. They were electrically neutral and therefore could be called neutrons.

If Chadwick were right, it would be the first known occurrence of neutrality on the subatomic level, but physicists seized upon the explanation eagerly. Not only did it explain N neatly and elegantly, but it also supplied a particle that had already been suggested as the only way of accounting for certain nuclear properties that until then had been puzzling physicists.

It became clear almost at once that atomic nuclei (all except that of the simplest hydrogen isotope, which was a simple proton) were made up of combinations of protons and neutrons and that it was by changing the nature of the combination through bombardment by subatomic particles that nuclear reactions were brought about.

Once neutrons were recognized and once methods for producing them were discovered, it was quickly understood that they offered a new and particularly exciting bombardment device.

Since neutrons were uncharged, they were not repelled by the positively charged atomic nuclei. If they happened to be aimed correctly, there was no repelling force to swerve them away or turn them back. The neutrons just moved on remorselessly and struck the nuclei.

The percentage of hits was therefore increased considerably if one used neutrons rather than protons or alpha particles. Even so, however, the percentage would remain extremely small so that the chance of getting out more energy than one was putting in still seemed out of the question.

The disadvantage of the situation was that there was no good way of accelerating neutrons. Electrically charged subatomic particles were accelerated by a properly manipulated electromagnetic field. The field acted upon the electric charge, which served as a "handle" for the particle. The uncharged neutron had no handle, so that if it were emitted from nuclei with a certain amount of energy, that was all the energy it could have. You could give it no more.

Since, as it seemed, the less energy a subatomic particle had, the less effective it would be in inducing a nuclear reaction, the advantage of the neutron's neutrality seemed to be balanced, and perhaps more than balanced, by the disadvantage.

Neutrons as produced were, of course, capable of inducing nuclear reactions. This was demonstrated in 1932, the very year of the neutron's discovery by, among others, the American chemist William Draper Harkins (1873–1951). Fairly energetic neutrons were used in these cases.

In 1934, however, the Italian physicist Enrico Fermi (1901–54) found that neutrons lost energy if they passed through materials made up of light atoms, such as water or paraffin.

What happened was this. If a neutron hits a massive atom, that neutron might be absorbed and induce a nuclear reaction; but it might also simply bounce. The massive atom is so massive that it hardly moves under the impact and the neutron bounces back at its original speed of approach—like a ball bouncing back from a wall. The neutron, in this way, keeps all its energy.

If a neutron, however, hits a relatively light nucleus and bounces, the light nucleus recoils somewhat and takes up some of the momentum so that the neutron bounces back with less speed and energy than it had approached. After several bounces of this sort, the neutron ends up with no more energy than ordinary atoms would have at that temperature. It would move very slowly indeed for a subatomic particle and it is then referred to as a "slow neutron."

One would suppose that slow neutrons, possessing virtually no energy, would be useless as far as inducing nuclear reactions were concerned, but this is *not so*.

Fermi made the crucial discovery that slow neutrons are *more* effective in inducing nuclear reactions than fast neutrons are. What happens is this:[1] Although electric repulsion (or attraction) is not a factor in the case of the uncharged neutrons, there are certain nuclear forces that actually *attract* a neutron if they get close enough to a nucleus, and would do so much more strongly than an electric charge would.

However, whereas an electric charge can make itself felt at a considerable distance, the nuclear attraction falls off so rapidly with distance that it will make itself felt only in the immediate neighborhood of a nucleus. Since a slow neutron is bound to remain near a nucleus longer than a fast one would, the slow neutron would have a greater chance of being sucked into the nucleus and of inducing a nuclear reaction.

Fermi began to use slow neutrons for bombardment and found that in many cases what happened was that the neutron was absorbed and added to the nucleus. The resultant nucleus, with the extra neutron, was usually radioactive and achieved stability by giving off an electron. This process changed a neutron to a proton, so that the final nucleus possessed one proton more than the original.

The chemical nature of an atom depends on the number of protons in the nucleus (the "atomic number"), so neutron bombardment frequently changed an atom of a particular element with a particular atomic number to an atom of

[1] The explanation arose out of the work of the Japanese physicist Hideki Yukawa (1907–   ) in 1935.

another element which was one higher up in the atomic-number scale.

For instance, if cadmium (atomic number 48) were bombarded with neutrons, indium (atomic number 49) would be formed.

Fermi at once thought of uranium, the element with the highest known atomic number, 92. What would happen if uranium were bombarded with slow neutrons?[2]

If the same thing happened to uranium that happened to other elements, a product one higher in atomic number would form and Fermi would have produced element 93. But element 93 did not occur in nature as far as was known, so that Fermi might in this way produce a new manmade element and that would be as sensational as discovering a new planet.

In 1934, Fermi began to bombard uranium with slow neutrons and, after a while, he decided that he might actually have succeeded in producing atoms of element 93. He was not certain of this. The results were not clear cut and there were evidences of radiation he could not explain. For that reason, Fermi would have held off making the announcement, but Italy's Fascist dictator, Benito Mussolini, anxious for a dramatic Italian scientific feat, forced a premature disclosure.

It was not altogether premature, to be sure. In 1940, after some of the confusing aspects of the nuclear reaction had been cleared up, two American physicists, Edwin Mattison McMillan (1907–   ) and Philip Hauge Abelson (1913–   ), showed that element 93 had actually been formed. Indeed, after a neutron had been added to uranium, and that neutron had been changed to a proton, a second neutron was eventually changed to a proton also to form element 94.

Since uranium had been named for the planet Uranus, the next two elements were named for Neptune and Pluto,

---

[2] Did you think I wouldn't get back to uranium?

the planets beyond Uranus. Element 93 became neptunium
and element 94 became plutonium.

In all this, Fermi wasn't thinking of the tapping of nuclear
energy. Even slow neutrons didn't strike often enough to
allow an adequate return of energy—not nearly.

Someone else, however, *was* thinking of nuclear energy.
He was Leo Szilard, a Hungarian-born physicist (1898–
1964). He had been teaching in Germany, but he was
Jewish, and when it looked as though Hitler was coming
to power, Szilard was wise enough to leave for Great Britain.

Szilard had been set to thinking about nuclear energy by
one of H. G. Wells' stories in which "atomic bombs" had
figured. It occurred to Szilard that if a nucleus absorbed a
neutron and underwent a nuclear reaction that liberated
two neutrons, each of those might induce a similar nuclear
reaction that would liberate a total of four neutrons, which
would in turn . . .

The initial investment of one neutron would, in other
words, set off a "chain reaction" that would produce a vast
quantity of energy. Chain reactions were well known in ordinary chemistry—anytime a small spark sets off a forest fire
or a dynamite explosion, we have an enormous example of a
chemical chain reaction. Why not a nuclear chain reaction,
then?

Szilard thought that such a nuclear chain reaction might
take place if the element beryllium were bombarded with
neutrons. I believe he even obtained a patent for a device
making use of this supposed nuclear reaction and assigned it
to the British Government. Unfortunately, the figures available for the beryllium nucleus were not quite accurate, and
when they were corrected the chance of a nuclear chain reaction involving beryllium disappeared.

Szilard then thought that the proper thing to do was to
bombard each element with neutrons with the intention of

seeing whether something would result in some specific case that would lead (with whatever necessary modification) to a nuclear chain reaction. For that, he needed money.

He approached the Russian-British biochemist Chaim Weizmann (1874–1952), who was also Jewish and who was impressed with the importance of the idea. Weizmann undertook to raise a few tens of thousands of dollars, but failed. No one was interested enough to invest.

Later on, Szilard decided that had been a very lucky failure. He and Weizmann were, of course, keenly aware of the danger of Nazism, as any Jews would be sure to be. They saw that the first and easiest (almost inevitable) use of nuclear energy would be the kind of atomic bomb H. G. Wells had talked about, and they knew that the Nazis must not get it first.

Well, then, if Szilard and Weizmann had started working on it in the middle nineteen-thirties and word had gotten out (as it surely would have), the western powers, anxious for peace and eager not to annoy the Nazis, would never have supported it. The Nazis, however, planning war, might well have begun a full-scale effort that would have gotten them the bomb first.

Clearly, Szilard could have been sure of western support only if a war with Germany was imminent or had actually begun. But I'm getting ahead of the story.

Fermi's announcement of element 93 carried very little conviction, as it happened. Other nuclear physicists tried to confirm the discovery and they ran into the same difficulties that Fermi himself had experienced. There were a number of different sets of subatomic particles of different energies being produced and the formation of element 93 simply could not account for them all. Other things must be happening, too.

One German chemist, Ida Tacke Noddack (1896–), a codiscoverer of the element rhenium, was openly skeptical

that any element 93 had been formed at all. She apparently
believed that uranium had the most complicated atoms capable of existing and that any major disturbance of the
nuclei of such atoms would simply cause them to split into
fragments, or undergo "fission" (from a Latin word for
"split"). She didn't use the word "fission," however, and she
had no evidence at all to back up her belief, so her suggestion was completely ignored.

Until then, all nuclear reactions had involved the emission of subatomic particles of comparatively small mass.
The most massive emitted particle was the alpha particle
with a mass of 4 on the atomic-weight scale. Physicists were
reluctant to move beyond this.

Two who were particularly engaged in trying to work out
the problem of what happened to uranium under neutron
bombardment were the German physicist Otto Hahn
(1879–1968) and his Austrian coworker, Lise Meitner
(1878–1968). Meitner was Jewish but she was an Austrian
national, so she could work in Nazi Germany without immediate danger during the early years of Hitler's ascendancy.

It occurred to Hahn and Meitner that a double dose of
alpha particle emission, might be brought about by neutron
bombardment, and that this would convert uranium atoms
into radium atoms. (I don't know the details of the reason-
ing and I sometimes wonder if Hahn and Meitner thought
of the double-alpha-particle emission—which, in hindsight,
seems so unlikely—because of the general glamour of radium. If so, it was radium's last moment of glamour.)

Hahn and Meitner could demonstrate this to be so if they
could detect tiny traces of radium in the neutron-bombarded uranium. However, so very few of the uranium
atoms would have undergone such a change that only a few
radium atoms would have been formed. How could so tiny
a trace of radium be detected?

Well, radium is an "alkaline earth metal"; that is, it is
chemically similar to the elements calcium, strontium and

barium. It is most similar to barium. In fact, it is virtually a chemical twin of barium, and if radium weren't radioactive, this fact of twinhood would be its most notable characteristic.

In that case, suppose one were to add barium to the neutron-bombarded uranium and force the uranium to undergo chemical reactions that would separate out of it the barium that had been added to it. Whatever would serve to separate the barium from the uranium (and the chemical methods for doing so were well known) would also serve to separate the radium from the uranium. Radium and barium are so similar chemically that what would work for one would work for the other.

The barium that was originally added would, of course, be perfectly stable and nonradioactive. The barium that would be separated would come out with the radium attached and would therefore seem radioactive. That in itself would be a good indirect sign that the Hahn-Meitner theory of double-alpha-particle emission was correct.

The next step would be to subject the barium-radium mixture to rather tedious and delicate chemical reactions that would separate the two. (Barium and radium are very similar in chemical properties but they are not entirely identical. They *can* be separated.)

Before all this could be carried through, Nazi Germany invaded and annexed Austria in March 1938 and Meitner's position in Berlin became untenable. She slipped across the border to the Netherlands and from there she went to Stockholm, Sweden. The Danish physicist Niels Bohr (1885–1962), a vigorous anti-Nazi, helped her get established in her new home.

Hahn continued his work with Fritz Strassman (1902–   ), however, and when the added barium was separated, it *did* emerge radioactive, which was cause for jubilation. However, the next step failed. Nothing they could do would separate the radium from the barium.

Hahn felt himself forced to what seemed a ridiculous conclusion. If the radium couldn't be separated from the barium, then it wasn't radium.

What, then, was the only substance that couldn't be separated from barium by any chemical means? Barium itself!

Could it be that when uranium was bombarded by neutrons it formed a radioactive isotope of barium? That when ordinary barium was added and then separated, the radioactive barium came out with it?

But the atomic number of uranium was 92 and the atomic number of barium was 56. If the latter was formed from the former, the uranium atom must have split into two nearly equal halves as a result of its absorption of a neutron. It would be a case of uranium fission; it would be Noddack's suggestion revived—but with the accompaniment of evidence.

Hahn thought about it but didn't quite dare go public with the suggestion of uranium fission. It seemed too outlandish.

Meanwhile, in Sweden, Meitner was coming to precisely the same conclusion, and she *did* decide to go public. Perhaps this was because the shattering blow of exile had made her less concerned with trivialities such as what people would think.

With the help of her nephew, Otto Robert Frisch, who worked in Bohr's laboratory, Meitner prepared a letter, dated January 16, 1939, outlining her suggestion of uranium fission and sent it to the British scientific journal *Nature* for publication.

As it happened, though, Frisch told his boss, Bohr, of the letter before it was published. Bohr was going to the United States to attend a physics conference in Washington, D.C., on January 26, 1939, and he spread the word there, also before the letter was published.

And thus it came about . . .

Because no one had listened to Noddack; because Hahn had hesitated; because Meitner was in exile; because her nephew worked for Bohr; because Bohr happened to be in the United States at the right time—it was not German physicists who followed up the first experimental evidence of uranium fission; but *American* physicists.

It's enough to make one burst into a cold sweat in retrospect as one envisions the mushroom cloud over New York and the swastika flag over the White House. And even so the narrow squeaks are not over, as we shall see in the next chapter.

12.

The Finger of God

In 1755, the British sent an army to North America, under General Edward Braddock, in order to dispute the French expansion into western Pennsylvania.

Braddock took a liking to a twenty-three-year-old Virginian who had already fought the French there (unsuccessfully) and appointed him an aide-de-camp. He was the only colonial aide among a group of British.

Braddock then marched his men toward the site of modern Pittsburgh and attempted to fight there in the style of European battles, with all his men carefully lined up and all firing a volley on order. They were opposed by French and Indians who, observing that they were fighting in a trackless forest, got each behind a tree.

The French and the Indians fired ad lib from behind those trees and mowed down the British, who made splendid marks in their bright-red uniforms. The British had nothing visible at which to fire in return, and when they tried to take cover, Braddock beat them back into line with yells, oaths and the flat of his sword.

The British were slaughtered, of course, and Braddock was fatally wounded, dying four days later muttering, "Who would have thought it?"

That any of the army was left alive at all was thanks to the Virginian aide-de-camp who, when the British finally broke and ran, covered their retreat by having his own Virginia troops fight Indian style.

The young Virginian went through the battle without a scratch. Two horses were killed under him. Four bullets ripped through his clothes without touching him. He was the *only* aide-de-camp who remained alive (let alone totally unharmed) in that shambles.

That Virginian's name (you're ahead of me, I know) was George Washington.

I first heard this story in class when I was about ten years old. The teacher (whom I will call Mr. Smith) got very emotional about it and told us that it was clearly the finger of God. Washington, he said, had been saved so that twenty years later he could lead the colonies to victory in the Revolutionary War and thus establish the United States of America.

I listened to that with the deepest skepticism. In the first place, it seemed to me that God wasn't an American and had to care for all people equally. If he were really efficient he would have figured out some way of accomplishing his purpose without a battle and thus have saved *everyone's* life. But then a sudden, staggering thought occurred to me and I raised my hand excitedly.

The teacher pointed to me and I said, "How can you say that was the finger of God, Mr. Smith? For all any of us know, someone was killed in that battle who, if he had

lived, would have been better than George Washington,
and who would have figured out a way to make us independent without a war."

At that Mr. Smith turned red. His eyes bulged, he
pointed a finger at me and shouted, "Are you trying to tell
me that anyone would have been better than George Washington?"

And I was ten years old and very frightened and backed
down hastily—but only on the outside. Inside my head, I
held the fort and was certain that describing something as
representing the finger of God was silly. In every conflict of
every sort, whether between individuals or nations, what
seems like the finger of God to the winner seems surely like
the devil's hoofprint to the loser.

And yet how tempting it is to play the "finger of God"
game. I ended the last essay on the discovery of uranium
fission by pointing out the series of lucky accidents that led
to the initial work on the process being done in the United
States rather than in Nazi Germany, and clearly felt such
relief at that that one might almost suppose I thought the
finger of God had something to do with it.

Well, I'm not through.

Consider the situation of Leo Szilard in 1939. As I explained, he had been thinking of the idea of a nuclear chain
reaction. His first attempt in that direction involved the interaction of a neutron with a beryllium nucleus in such a
way that two neutrons were liberated. However, it took a
fast energetic neutron to interact with the beryllium nucleus, and only slow neutrons were liberated, neutrons with
too little energy to interact with further beryllium nuclei.

Uranium, on the other hand, undergoes fission when stimulated by *slow* neutrons. To be sure, it liberates fast neutrons in the process which are actually not as efficient in
breaking down uranium nuclei as slow neutrons are. (They
go too fast and don't linger in the neighborhood of a nucleus long enough for a good chance of reaction.)

However, while slow neutrons can't be hastened, fast neutrons can easily be slowed. If, then, you start fissioning uranium, and slow the neutrons produced, you can keep on fissioning uranium in rapidly accelerating fashion to produce a bomb of unprecedented and devastating power.

In 1939, it was plain to Szilard that the world was on the brink of war, that Nazi Germany might win such a war and that that nation represented a dire peril to civilization.

Szilard was further sure that it was quite possible for a uranium fission bomb to be developed in the course of the war, and it seemed plain that whichever side developed and used the bomb first would win the war, even if it happened to be on the brink of defeat before the use. Who, then, would get the nuclear bomb first, Germany or the United States? (There was an outside chance that Great Britain or France might. No one, at that time, would have felt that the Soviet Union or Japan had even an outside chance.)

Actually, Szilard may well have felt the odds were on the Germans for a number of reasons:

1. The scientific tradition in Germany was much stronger than that in the United States. Through the period from 1850 to 1914, Germany had led the world in scientific research, while the United States was so backward in this respect that any American craving a scientific career was almost bound to go to Germany for at least part of his graduate work. Germany was strong, specifically, in nuclear physics, and it was in Germany that the evidence for uranium fission had first been gathered.

2. Germany was under the absolute control of Adolf Hitler, who, if he became interested in the possibilities of a nuclear bomb, could, without hindrance, throw the entire resources of the nation behind its development, with money no object. The United States, on the other hand, was a democracy run by people for whom the most world-shaking goal was re-election. To put a lot of money into some fly-by-night science fiction scheme might risk a congressional seat, heaven forbid.

3. Germany was a closed society and if Hitler grew interested in the possibility of a nuclear bomb, any German discoveries in that direction would have been kept in deepest secrecy. In the United States, however, all discoveries would be promptly published and discussed so that Germany would benefit from any advance Americans made—but not vice versa.

Szilard felt it was up to him to do something about this and to shift the odds, as far as he could, in favor of the United States.

I have listed the three points in order of decreasing intractability. The first point, for instance, Germany's scientific tradition and the United States' lack of one, is a historical fact and nothing can be done about it—except that it was changing, and I imagine Szilard was aware of that.

Since World War I, Germany had been losing its preeminence in science and the United States had been gaining rapidly. Furthermore, Hitler himself was Szilard's best ally in this respect. Hitler's paranoid racial views had greatly weakened German science and had flooded the West with scientist-refugees who had the ability to devise a nuclear bomb for the United States and the strongest possible motive to do so.

In fact, one might imagine, as an "if" of history, a Hitler who differed from the real one in not being obsessed with "racial purity." If so, those whom he drove out of Germany in the name of such "purity" would have remained in place. There is no reason to suppose they would not have been routinely patriotic Germans and they might then have contributed mightily to the building of a nuclear bomb for Germany rather than for the United States, and Germany might now be the dominant nation on the planet.

We might say "How ironic!" and raise our hands in amazement at the way Hitler defeated himself, and talk of the finger of God, except that this is not an unprecedented sort of event. It happened at least twice before in European history in just as spectacular a fashion. Spain under Philip

III evicted the Moriscos (Christians of Moorish descent) and France under Louis XIV evicted the Huguenots (Christians of Protestant persuasion). In each case the nation that did the evicting in the name of religious "purity" lost a particularly valuable part of the population, weakened itself permanently and strengthened its enemies in proportion.

Has humanity learned a lesson from this? Of course not. Right now, Vietnam is laboring to evict Vietnamese of Chinese descent and there is an absolute certainty that Vietnam will be permanently weakened as a result.

It doesn't take the finger of God to make human beings place their prejudices ahead of their good sense. I'd be more likely to be tempted to believe it in the reverse case.

But back to Szilard. He could scarcely gamble on Hitler's having weakened German science sufficiently to make the situation safe, so he had to tackle points 2 and 3.

He began a one-man letter-writing campaign pointing out the possibilities of a nuclear bomb and asking scientists in the field to keep their work secret. It was hard for scientists to agree to this. Free and open communication among scientists, together with complete and early publication, is the very foundation of scientific progress.

And yet the case was unprecedented and little by little Szilard won out. By April 1940, there was a voluntary system of self-censorship on the subject and public discussion of nuclear fission ceased. Szilard had taken care of point 3 and that meant that Germany could no longer count on our being kind enough to help her destroy us.

By then, however, it began to seem as though Germany didn't need our help. By April 1940, Hitler had come to an agreement with the Soviet Union, begun the war, destroyed Poland, taken over Denmark and Norway, all of this while Britain and France remained in a state of paralysis. Shortly after Szilard's victory, Hitler took France and began to subject Great Britain to a merciless air bombardment. And in

1941, he turned on the Soviet Union, after clearing out the Balkans, and bit deep into the Russian homeland.

It looked as though he would have all Europe, and perhaps eventually all the world, *without* nuclear weapons.

Now it became important for the United States to develop a nuclear bomb not only in order that we might have it ahead of the Germans but perhaps as a last-ditch defense against otherwise inevitable defeat. And we only had a few years to do it in.

It is hard now, for those who didn't live through it as I did, to understand the desperation of those days. It was quite possible for the United States to fritter away its time and chances while Germany charged ahead to work out and make use of unprecedented weapons.

Consider the case of rocketry, for instance. Modern rocketry began in the United States with Robert Goddard in 1926, but Goddard remained a one-man operation. The government would not help out. It is doubtful whether in the twenty years between 1926 and 1946 there would have been a single Congressman with the vision to support rocketry or principled enough to risk re-election over it.

That was not the case in Germany, where government support of rocketry began early in the game so that, by 1944, V-2 missiles were bombarding Great Britain.

With this in mind, we can again wonder over the fact that Germany didn't win, and again it was a case of Hitler's defeating himself. For one thing, his interest in rockets and missiles drowned out his interest in the nuclear bomb. In the war emergency, he seemed to have room inside himself for only one secret weapon at a time.

More fundamentally, Hitler's desire to send his troops goose-stepping across Europe while he was still alive and young enough to enjoy the destruction led him to a premature war. I suspect he didn't want to build up a war machine that some successor would then use to conquer the world.

There were, after all, historical precedents for that, and Hitler, an ardent student of history, knew about it. Philip of Macedon built up an army that his son, Alexander, used to conquer all the Persian Empire and it is the son who is called "the Great."

Closer to home, Frederick William I of Prussia built up a beautifully polished army which his son, Frederick II, used to defeat Austrian and French armies, and it is the son who is called "the Great."

Presumably Hitler wanted to be Philip and Alexander combined, and he didn't want to risk waiting too long.

He was, however, still only fifty years old in 1939, and he might have risked waiting for, say, five more years. If he had, he could have been sure that the western powers would have utterly wasted the time. Great Britain and France would have been pleased that Germany was making no more territorial demands after Munich and would have leaned over backward to avoid irritating Hitler. Franklin Roosevelt would not have run for a third term in 1940 if the world were at peace, or would have been defeated if he had tried, and his successor, whoever he was, would have been less able to withstand isolationist sentiment in the United States.

Hitler could then have mounted major programs to develop both missiles and the nuclear bomb, with no competition whatever from the West. The Soviet Union would also be working in both directions, I'm sure, but I suspect that Hitler would have gotten there first.

Then in 1944 or 1945, Hitler would have had missiles and nuclear bombs ready or almost ready for rapid production and improvement as necessary. He could have started the war and reserved his secret weapons for emergencies. If the war went unexpectedly badly or endured too long or if it looked as though, behind her insulating oceans, the United States might catch up with and surpass Germany in the production of conventional weapons, two or three nuclear bombs exploded over American cities by missile from some submarine offshore would, I think, have been enough to end it all and Hitler would rule the world.

But it didn't happen. Hitler, without the benefit of hindsight, may not have seen all this, but my feeling is that none of these possibilities would have interested him. He simply wouldn't wait longer because he wouldn't take the risk of losing the credit of conquest and so he lost his chance by only tha-a-at much.

The finger of God? Why? Surely it doesn't require the forces of heaven to make a paranoid egomaniac act like a paranoid egomaniac.

But Szilard couldn't count on all this. He couldn't foresee the future and he couldn't be sure that Hitler had been premature. It certainly didn't look as if he were in 1941.

No! The United States had to have the nuclear bomb and there was no way it could do that without a massive government program, and an expensive one, to support the necessary research and engineering. But how on earth could the government be persuaded to invest the money? Congress? Forget it! With the world burning up on every side, the House of Representatives renewed the draft by *one* vote. One congressman opposed renewal by saying that if there were an invasion, all Americans "would spring to arms." He didn't say what arms or how they would be trained to use them.

It would be much better to try President Roosevelt, but he was only the President and he would surely be eviscerated by Congress and the people if he spent a lot of money on something that wasn't of immediate and visible use to some large voting section of the public. In order to get around that, Roosevelt would have to be impressed with the urgency of the situation—so impressed that he would risk political suicide.

How the devil could Roosevelt be impressed to that point? It was a scientific matter, to be sure, but it sounded like science fiction, and there is nothing that infuriates the down-to-earth boneheads of the world like something that sounds like science fiction. To get rid of the taint, the matter would have to be presented by some scientist so towering in reputation that no one would question his statements.

There was only one living scientist who was an absolute legend to the world—even to those who knew nothing about science except that two and two added up to something between three and five. That was Albert Einstein.

Szilard therefore enlisted the help of two friends, Eugene Paul Wigner and Edward Teller. All three were brilliant nuclear physicists of Hungarian birth who had fled Hitler. All three were perfectly convinced of the dangers the world faced and the need to have the bomb in view of the Nazi menace. And all three went to see Einstein, who had also been a victim of the Nazis.

It was not easy to persuade Einstein to put his name to the letter. He was a convinced pacifist and he did not desire to put this dreadful weapon in the hands of human beings, but he could see the dangers and the incredible dilemma the world faced. It was hell both ways, but he had to make a choice, and he put his name to the letter that Szilard had written for him to sign.

The letter went to Roosevelt and the use of the Einstein name apparently supplied the necessary clout. Roosevelt decided to go for broke and to authorize a secret project for the development of the nuclear bomb, one that was eventually to cost $2 billion. (One can imagine the ridicule that would have been heaped on Roosevelt's head by the Proxmires of the world, if the project had failed.)

Even a presidential decision must go through red tape, however, and it wasn't until a particular Saturday late in the year that Roosevelt finally signed the order that set up what came to be called the Manhattan Project—a deliberately meaningless name designed to mask its real purpose.

That, as it happened, was an incredibly close call. It is a

good old American custom, after all, not to do anything important on the weekend and even presidents are American sometimes.

If Roosevelt had indeed delayed the signing till Monday, who knows when it would have been signed, or if it would have been signed at all.

The day on which the order was signed was Saturday, December 6, 1941, and the next day was Sunday, December 7, 1941—the day on which Japanese warplanes bombed Pearl Harbor. After that there was nothing but chaos in Washington for quite a while.

The order was, however, signed on the last possible day (the finger of God? the devil's hoofprint?), and the nuclear bomb was developed, and the United States had it first.

Szilard had won.

And yet it turned out we didn't need the bomb after all. Hitler's Germany never developed an atom bomb, and it developed its missiles too late in the war to win it.

On April 30, 1945, Hitler died a suicide and on May 8, 1945, Germany surrendered. The world could relax.

To be sure, Japan was still fighting, but its fleet was gone, its armies had been defeated, its cities were smashed into rubble. It was on the point of surrender.

Many of the scientists who had been anxious to devise a nuclear bomb now no longer felt one to be necessary. As long as it was a matter of getting one before the Nazis did, or of getting one to prevent our final defeat, then we had to have one. The horror of the bomb seemed, at the time, to be preferable to the horror of a Nazified world.

But once Nazi Germany was destroyed and Japan was clearly on the point of defeat, why not stop work on the bomb, hold it in reserve for future emergencies or reveal what work had been done and place it all under international control—or *something, anything* to avoid what might, and did, come to pass; a world with opposing powers armed from end to end with nuclear weapons and with world destruction an always imminent possibility?

And yet the development of the nuclear bomb went on. On July 16, 1945, the first nuclear bomb explosion in the history of the world went off in Alamogordo, New Mexico. On August 6, 1945, the second nuclear bomb explosion took place over Hiroshima, Japan, and on August 9, 1945, the third took place over Nagasaki. The Japanese formally surrendered on September 2.

Why? One might defend the Alamogordo explosion. After all, the work and the investment had been huge and there was an overwhelming curiosity to see if the bomb worked.

But then why use it on a dying enemy?

The reasons advanced after the event were that the diehard, fanatic Japanese would never surrender unless and until the Americans actually invaded the Japanese home islands and that the Japanese would then fight with incredible ferocity, causing the deaths of at least 100,000 Americans, let us say, and 500,000 Japanese. To bomb two cities instead would represent a net saving of hundreds of thousands of American *and* Japanese lives and would therefore be a great humanitarian act.

I didn't believe that at the time, and I don't believe it now.

However, the Japanese weren't the real enemy at that moment. The real enemy was our ally, the Soviet Union.

At the Yalta Conference, held in February 1945, the Soviet Union had promised to declare war on Japan three months after the Nazi surrender, for they needed that much time to transfer supplies and men across five thousand miles from the western borders of the Soviet Union to the eastern borders. This was agreed to.

Despite all the facile talk about how you can't trust the Soviet Union, the fact is that the Soviet Union generally lives up to the letter of specific agreements. (It may violate the spirit, but that's another thing.) If it said three months, it meant three months, and three months after May 8, 1945 is August 8, 1945. On that day, in fact, the Soviet Union declared war on Japan.

The United States, however, had been fighting Japan for three and a half years. It was a bitter fight and we had the humiliation of Pearl Harbor to avenge. We wanted to be sure we got the full credit of the victory. If Japan surrendered some time after the Soviet army pushed into Manchuria it might look as though that was the crowning blow and we might lose the credit. We therefore hastened like mad to get at least one nuclear bomb ready to drop on a Japanese city *before* the Soviets came in, and we made it by two days. After that, the Soviet entry was only a detail and the whole world knew who had defeated the Japanese. The United States had.

What's more, we knew very well that we were going to be competing with the Soviet Union for influence in Europe and the world once the war was over and we decided that it was necessary for the Soviet Union to know that we had this terrible weapon. What's more, we had to do more than talk about it or hold empty demonstrations over desert or sea. It had to be used on a *city* so that the death and destruction it caused could be plainly seen. So we had to do it quickly, before the Japanese surrendered and deprived us of an enemy to do it to. Hiroshima and Nagasaki were coldblooded demonstrations intended for the Soviet Union. At least that's the way I see it.

It is possible to argue that this alone prevented a Soviet-American war in the years after World War II. Since by preventing this, millions of lives were saved, the nuclear bombing of the Japanese cities could again be hailed as a humanitarian act. It is further possible to argue that managing to get the bomb in time to get the bombing done, just before the Japanese surrender would have made it impossible, is another example of the finger of God.

On the other hand, might it not be possible to argue that the narrow margin that permitted us to develop and use the nuclear bomb at the end of World War II imbued the United States with a feeling of overconfidence that kept it from attempting conciliation with the Soviet Union at a time when the Soviet Union was sufficiently weak from its battering by the Germans to welcome such conciliation?

Might it not be possible to argue further that the over-confidence led us into a series of foreign policy mistakes for which we are paying now?

The finger of God? The devil's hoofprint?

Or perhaps we should stop looking for supernatural causes and take a close hard look at human folly. I don't think we need anything more than that.

13.

Clone, Clone of My Own

On December 12, 1968, I gave a talk to a meeting of doctors
and lawyers in San Jose, California.[1] Naturally, I was asked
to speak on some subject that would interest both groups.
Some instinct told me that medical malpractice suits might
interest both but would nevertheless not be a useful topic. I
spoke on genetic engineering instead, therefore, and, toward the end, discussed the matter of cloning.

In the audience was my good friend of three decades, the
well-known science fiction writer, bon vivant and wit Ran-

[1] Those of my Gentle Readers who know that under no circumstances will I take a plane need not register shock. I traveled to California and back by train. Yes, they still run.

dall Garrett. Out of the corner of my eye I noticed a piece
of paper placed on the podium as I talked about cloning. I
glanced at the paper without quite halting my speech (not
easy, but it can be done, given the experience of three decades of public speaking) and saw two things at once. First,
it was one of Randall's superlative pieces of satiric verse,
and second it was clearly intended to be sung to the tune of
"Home on the Range."

Needed to understand the verse is merely the fact that,
genetically, the distinction between human male and female
is that every male cell has an X- and a Y-chromosome and
that every female cell has two X-chromosomes.[2] Therefore,
if, at the moment of conception or shortly thereafter, a Y-
chromosome can somehow be changed to an X-chromosome,
a male will *ipso facto* be changed into a female.

Here, then, is Randall's Song, to which I took the liberty
of adding four more verses myself:

(*1st verse*)

O give me a clone
Of my own flesh and bone
With its Y-chromosome changed to X;
And when it is grown
Then my own little clone
Will be of the opposite sex.

(*chorus*)

Clone, clone of my own,
With its Y-chromosome changed to X;
And when I'm alone
With my own little clone
We will both think of nothing but sex.

[2] See COUNTING CHROMOSOMES, *F & SF*, June, 1968.

(*2nd verse*)

O give me a clone,
Hear my sorrowful moan,
Just a clone that is wholly my own;
And if it's an X
Of the feminine sex
O what fun we will have when we're prone.

(*3rd verse*)

My heart's not of stone
As I've frequently shown
When alone with my dear little X;
And after we've dined,
I am sure we will find
Better incest than Oedipus Rex.

(*4th verse*)

Why should such sex vex
Or disturb or perplex
Or induce a disparaging tone?
After all, don't you see,
Since we're both of us me,
When we're making love I'm alone.

(*5th verse*)

And after I'm done
She will still have her fun,
For I'll clone myself twice ere I die.
And this time, without fail,
They'll be both of them male,
And they'll each ravish her by-and-by.

When I was through with my talk and with the question-
and-answer session, I sang Randall's Song in my most reso-
nant baritone and absolutely brought the house down.

Three and a half weeks later I sang it again at the annual banquet of the Baker Street Irregulars, that fine group of Sherlock Holmes fanciers, adjusting it slightly to its new task (*O give me some clones/Of the great Sherlock Holmes/With their Y-chromosomes* . . .) and brought the house down again.

But you may, by now, be asking yourself, "What's a clone?"

It's been in the news a great deal lately, but recognizing a word, and knowing what it represents can be two different things. So let's go into the matter . . .

The word "clone" is Greek, exactly as it stands, provided you spell it in Greek letters, and it means "twig."

A clone is any organism or group of organisms that arises out of a cell or group of cells by means other than sexual reproduction. Put another way, it is an organism that is the product of asexual reproduction. Put still another way, it is an organism with a single parent, whereas an organism that arises from sexual reproduction (except where self-fertilization is possible) has two parents.

Asexual reproduction is a matter of course among one-celled organisms (though sexual reproduction can also take place) and it is also very common in the plant world.

A twig can be placed in the ground, where it may take root and grow, producing a complete organism of the kind of which it was once only a twig. Or the twig can be grafted to the branch of another tree (of a different variety even), where it can grow and flourish. In either case, it is an organism with a single parent and sex has had nothing to do with its making. It is because human beings first encountered this asexual form of reproduction in connection with fruit trees probably, that such a one-parent organism of nonsexual origin is called a "twig"—that is, "clone."

And what of multicellular animals?

Asexual reproduction can take place among them as well. The more primitive the animal, that is, the less diversified and specialized its cells are, the more likely it is that asexual reproduction can take place.

A sponge, or a freshwater hydra, or a flatworm, or a starfish, can, any of them, be torn into parts and these parts, if kept in their usual environment, will each grow into a complete organism. The new organisms are clones.

Even organisms as complex as insects can in some cases give birth to parthenogenetic young, and, in the case of aphids, for instance, do so as a matter of course. In these cases, an egg cell, containing only a half set of chromosomes, does not require union with a sperm cell to supply the other half set. Instead, the egg cell's half set merely duplicates itself, producing a full set, all from the female parent, and the egg then proceeds to divide and become an independent organism, again a kind of clone.

In general, though, complex animals and, in particular, vertebrates, do not clone but engage in sexual reproduction exclusively.

Why? Two reasons.

In the first place, as an organism becomes more complex and specialized, its organs, tissues and cells become more complex and specialized as well. The cells are so well adapted to perform their highly specialized functions that they can no longer divide and differentiate as the original egg cells did.[3]

This seems a terrible disadvantage. Organisms that can clone, reproducing themselves asexually, would seem to be much better off than other organisms—who must go to the trouble of finding partners and who must engage in all the complex phenomena, both physical and chemical, involved in sexual reproduction. Think of all the human beings who, for one slight flaw or another, can't have children—a prob-

---

[3] This is not mysterious. We see an analogy on the social plane. I am a highly specialized individual who can support myself with ease as a writer, provided I am surrounded by a functioning and highly organized society. Place me on a desert island and I shall quickly perish since I don't know the first thing about the simplest requirements for self-support.

lem that would be unknown if we could just release a toe
and have it grow into another individual while we grew another toe.

Here comes the second reason, then. There's an evolutionary advantage to sexual reproduction that more than
makes up for all the inconveniences.[4] In cloning, the genetic
contents of new organisms remain identical with those of
the original organisms, except for occasional mutations. If
the organism is very efficiently adapted to its surroundings,
this is useful, but it is an extremely conservative mechanism
that reduces the chance of change. Any alteration in the environment could quickly lead to the extinction of a species.

In the case of sexual reproduction, every new organism
has a brand-new mix of genes, half from one parent, half
from another. Change is inevitable, variation from individual to individual is certain. A species in which sexual reproduction is the norm has the capacity to adapt readily to
slight alterations in environment since some of its variants
are then favored over others. Indeed, a species can, through
sexual reproduction, split with relative ease into two or
more species that will take advantage of somewhat different
niches in the environment.

In short, a sexually reproducing species evolves much
more quickly than a cloning species and such difficult-to-
evolve specializations as intelligence are not likely to arise
in the entire lifetime of a habitable planet, without sexual
reproduction.

Yet in one specialized way cloning can take place in even
the most advanced animals—even in the human being.

Consider a human egg cell, fertilized by a human sperm

---

[4] Please don't write to tell me that the activities involved in sexual
reproduction are not inconvenient at all, but are a lot of fun. I know
that better than you do, whoever you are. The fun is an evolutionarily
developed bribe designed to have us overlook and forgive the inconveniences. If you are a woman, you will see the point more quickly,
perhaps, than a man will.

cell. We now have a fertilized egg cell which contains a half set of genes from its mother and a half set from its father.

This fertilized egg cell cannot become an independently living organism for some nine months, for it must divide and redivide within its mother's womb, and be nourished by way of its mother's bloodstream. It must develop, specialize and grow larger until it has developed the necessary ability to live independently. Even after it emerges from its mother's womb, it requires constant and unremitting care for a period of time before it can be trusted to care for itself.

Nevertheless, the matter of necessary care is genetically irrelevant. The fertilized egg is already a separate organism with its genetic characteristics fixed and unique.

The first step in the development of the fertilized egg is that it divides into two cells, which cling together. Each of these two cells divides again, and each of the four that result divides again and so on.

If, after the first cell division, the two offspring-cells, for any reason, should happen to fall apart, each offspring-cell may then go on to develop into a complete organism of its own. The result is a pair of identical twins, each with the same genetic equipment and each of the same sex, of course. In a sense, each twin is a clone of the other.

There is no reason to suppose that this separation of offspring cells can't happen over and over, so that three or four or any number of organisms might develop from the original fertilized egg. As a matter of practical fact, however, a mother's womb can only hold so much and if there are multiple organisms developing, each is sure to be smaller than a single organism. The more organisms that develop, the smaller each one and, in the end, they will be too small to survive after delivery.

There are such things as identical triplets and quadruplets, but I doubt that any higher number of infants would survive long after birth without the advantages of modern medical technique. Even then it is hard enough.

Identical twins are very like each other and often display mirror-image characteristics. (I once had a chemistry professor with his nose canted to the left. His identical-twin brother had his nose canted to the right, I was told.)

It is also possible, however, though not usual, for a woman to bring two different egg cells to fruition at the same time. If both are fertilized, two children will be born who are each possessed of genetic equipment different from the other. What results is "fraternal twins," which need not be of the same sex and which need not resemble each other any more than siblings usually do.

Consider the fertilized egg again. Every time it divides and redivides, the new cells that form inherit the same genetic equipment possessed by the original fertilized egg.

Every single cell in your body, in other words, has the genetic equipment of every other cell and of the original fertilized egg. Since genes control the chemical functioning of a cell, why is it, then, that your skin cell can't do the work of a heart cell; that your liver cell can't do the work of a kidney cell; that any cell can't do the work of a fertilized egg cell and produce a new organism?

The answer is that though all the genes are there in every cell of your body, they aren't all working alike. The cell is an intricate assemblage of chemical reactions, chemical building blocks, chemical products and physical structures, all of which influence one another. Some genes are inhibited, some are stimulated, in a variety of ways depending on subtle factors, with the result that different cells in your body have genetic equipment in which only characteristic parts are working at characteristic rates.

Such specialized development begins in the earliest embryo, as some cells come into being on the outside of the embryo, some on the inside; some with more of the original yolk, some with less; some with first chance at absorbing nutrients from the maternal bloodstream, some with only a later chance. The details are clearly of the greatest importance to human biology, and biologists just don't yet know them.

Naturally, the ordinary "somatic cells" of an adult human body, with their genetic equipment working only in highly specialized ways, cannot divide into a whole organism if left to themselves. Many body cells, such as those of the muscles or nerves, have become so specialized they can't divide at all. Only the sex cells, eggs and sperm, retain the lack of genetic specialization required to produce a new organism under the proper circumstances.

Is there any way of unspecializing the genetic structure of somatic cells so as to allow them to develop into a new organism?

Well, the genes are contained in the nucleus of the cell, which makes up a small portion of the total and is marked off by a membrane of its own. Outside the nucleus is the cytoplasm of a cell and it is the material in the cytoplasm that provides the various chemicals that help serve to inhibit or stimulate the action of the genes.

Suppose, then, the nucleus of a somatic cell were surrounded with the cytoplasm of an egg cell. Would the genetic equipment in the nucleus unblock, and would the egg cell then proceed to divide and redivide? Would it go on to form an individual with the genetic equipment of the original somatic cell, and therefore of the person from whom the somatic cell was taken? If so, the new organism would be a clone of the person who donated the somatic cell.

The technique has been tried on different animals. You begin with an unfertilized egg cell and treat it in such a way as to remove its nucleus, either by delicately cutting it out or by using some chemical process. In the place of the removed egg cell nucleus, you insert the nucleus of a somatic cell of the same (or, possibly, an allied) species, and then let nature take its course.

This has been successfully tried with animals as complex as a tadpole.

It stops being easy after the frog, though. Frog eggs are

naked and can be manipulated easily. They develop in water and can just lie there after the micro-operation.

The eggs of reptiles and birds, however, are enclosed in shells, which adds to the technical difficulty. The eggs of mammals are very small, very delicate, very easily damaged. Furthermore, even if a mammalian egg has had its nucleus replaced, it would then have to be implanted into the womb of a female and allowed to come to term there.

The practical problems of mammalian cloning are such that there is no chance of its happening for some time yet.[5] Yet biologists are anxious to perform the feat and are trying hard. Eventually, they will no doubt succeed. What purpose will it serve?

If clones can be produced wholesale, a biologist can have a whole group of animals with identical genetic equipment —a set of ten thousand identical-twin mice, let us say. There are many animal experiments that can be conducted with the hope of more useful results if the question of genetic variation could be eliminated.

By the addition of other genetic-engineering techniques, it might be possible to produce a whole series of animals with identical genetic equipment, except that in each case, one gene is removed or altered—a different gene in each individual perhaps. The science of genetics would then advance in seven-league strides.

There would be practical uses, too. A prize bull or a champion egg-laying hen could be cloned and the genetic characteristics that make the record-breaking aspects of the animal possible would be preserved without the chance of diminution by the interplay of genes obtained from a second parent.

In addition, endangered species could have their chances of survival increased if both males and females could be

[5] Nevertheless, since this article was first published, it has happened. Mice have been cloned.

cloned over and over. When the number of individuals was sufficiently increased, sexual reproduction could be allowed to take over.

We might even dream of finding a frozen mammoth with some cell nuclei not entirely dead. We might then clone one by way of an elephant's womb. If we could find a male and a female mammoth . . .

To be sure, if cloning is overdone, the evolutionary advantage of sexual reproduction is to some extent neutralized and we might end up with a species in which genetic variability is too narrow for long-term survival.

It is important to remember that the most important genetic possession of any species is not this gene or that, *but the whole mixed bag*. The greater the variety of genes available to a species, the more secure it is against the vicissitudes of fortune. The existence of congenital disorders and gene deficiencies is the price paid for the advantage of variety and versatility.

And what about cloned human beings, which is, after all, the subject matter of Randall's Song?

These may never be as important as you think. The prospect of importance rests chiefly on certain misapprehensions on the part of the public. Some people, for instance, pant for clones because they think them the gateway to personal immortality. That is quite wrong.

Your clone is not *you*. Your clone is your twin brother (or sister) and is no more you than your ordinary identical twin would be. Your clone does not have your consciousness, and if *you* die, you are *dead*. You do not live on in your clone. Once that is understood, I suspect that much of the interest in clones will disappear.

Some people fear clones, on the other hand, because they imagine that morons will be cloned in order to make it possible to build up a great army of cannon fodder that despots will use for world conquest.

Why bother? There has never been any difficulty in

finding cannon fodder anywhere in the world, even without cloning, and the ordinary process of supplying new soldiers for despots is infinitely cheaper than cloning.

More reasonably, it could be argued that the clone of a great human being would retain his genetic equipment and would therefore be another great human being of the same kind. In that case, the chief use of cloning would be to reproduce genius.

That, I think, would be a waste of time. We are not necessarily going to breed thousands of transcendent geniuses out of an Einstein, or thousands of diabolical villains out of a Hitler.

After all, a human being is more than his or her genes. Your clone is the result of your nucleus being placed into a foreign egg cell and the foreign cytoplasm in that egg cell will surely have an effect on the development of the clone. The egg will have to be implanted into a foreign womb and that, too, will have an influence on the development of the organism.

Even if a woman were to have one of her somatic nuclei implanted into one of her own egg cells and if she were then to have the egg cell implanted into the womb of her own mother (who, we will assume, is still capable of bearing a child), the new organism will be born into different circumstances and that would have an effect on its personality, too.

For instance, suppose you wanted one hundred Isaac Asimovs so that the supply of F & SF essays would never run out. You would then have to ask what it was that made me the kind of writer I am—or a writer at all. Was it only my genes?

I was brought up in a candy store under a father of the old school who, although he was Jewish, was the living embodiment of the Protestant ethic. My nose was kept to the grindstone until I could no longer remove it. Furthermore, I was brought up during the Great Depression and had to find a way of making a living—or I would inherit the candy store, which I desperately didn't want to do. Furthermore, I lived in a time when science fiction magazines, and pulp magazines generally, were going strong, and when a young man could sell clumsily written stories because the demand was greater than the supply.

Put it all together, it spells M-E.

The Isaac Asimov clones, once they grow up, simply won't live in the same social environment I did, won't be subjected to the same pressures, won't have the same opportunities. What's more, when I wrote, I just wrote; no one expected anything particular from me. When my clones write, their products will always be compared to the Grand Original and that would discourage and wipe out anyone.

The end result will be that though my clones, or some of them, might turn out to be valuable citizens of one kind or another, it would be very unlikely that any one of them would be another Isaac Asimov, and their production would not be worthwhile. Whatever good they might do would not be worth the reduction they would represent in the total gene variability of humanity.

Yet cloning would not be totally useless, either. There would be the purely theoretical advantage of studying the development of embryos with known variations in their genes which, except for those variations, would have identical genetic equipment. (This would raise serious ethical questions, as all human experimentation does, but that is not the issue at the moment.)

Then, too, suppose it were possible to learn enough about human embryonic development to guide embryos into all sorts of specialized by-paths that would produce a kind of monster that had a full-sized heart with all else vestigial; or a full-sized kidney, or lung, or liver, or leg. With just one organ developing, techniques of forced growth (in the laboratory, of course, and not in a human womb) might make development to full size a matter of months only.

We can therefore imagine that at birth every human individual will have scrapings taken from his little toe, thus attaining a few hundred living cells that can be at once frozen for possible eventual use. (This is done at birth, because the younger the cell, the more efficiently it is likely to clone.)

These cells could serve as potential organ banks for the future. If the time were to come when an adult found he had a limping heart, or fading pancreas, or whatever; or if a leg had been lost in an accident or had had to be amputated; then those long-frozen cells would be defrosted and put into action.

An organ replacement would be grown and since it would have precisely the same genetic equipment as the old, the body would not reject it. Surely that is the best possible application of cloning.

THE
SCIENTISTS

14.

Alas, All Human

When I was doing my doctoral research back in medieval times, I was introduced to an innovation. My research professor, Charles R. Dawson, had established a new kind of data notebook that one could obtain at the university bookstore for a sizable supply of coin of the realm.

It was made up of duplicate numbered pages. Of each pair, one was white and firmly sewn into the binding, while the other was yellow, and was perforated near the binding so that it could be neatly removed.

You placed a piece of carbon paper between the white and yellow when you recorded your experimental data and, at the end of each day, you zipped out the duplicate pages and handed it in to Dawson. Once a week or so, he went over the pages with you in detail.

This practice occasioned me periodic embarrassment, for the fact is, Gentle Reader, that in the laboratory I am simply not deft. I lack manual dexterity. When I am around, test tubes drop and reagents refuse to perform their accustomed tasks. This was one of the several reasons that made it easy for me, in the fullness of time, to choose a career of writing over one of research.

When I began my research work, one of my first tasks was to learn the experimental techniques involved in the various investigations our group was conducting. I made a number of observations under changing conditions and then plotted the results on graph paper. In theory, those values ought to have fallen on a smooth curve. In actual fact, the values scattered over the graph paper as though they had been fired at it out of a shotgun. I drew the theoretical curve through the mess, labeled it "shotgun curve" and handed in the carbon.

My professor smiled when I handed in the sheet and I assured him I would do better with time.

I did—somewhat. Came the war, though, and it was four years before I returned to the lab. And there was Professor Dawson, who had saved my shotgun curve to show people.

I said, "Gee, Professor Dawson, you shouldn't make fun of me like that."

And he said, very seriously, "I'm not making fun of you, Isaac. I'm boasting about your integrity."

That puzzled me but I didn't let on. I just said, "Thank you," and left.

Thereafter, I would sometimes try to puzzle out what he had meant. He had deliberately set up the duplicate-page system so that he could keep track of exactly what we did each day and if my experimental technique turned out to be hopelessly amateurish, I had no choice but to reveal that fact to him on the carbon.

And then one day, nine years after I had obtained my Ph.D., I thought about it and it suddenly occurred to me that there had been no necessity to record my data directly in my notebook. I could have kept the data on any scrap of paper and then *transferred* the observations, neatly and in good order, to the duplicate pages. I could, in that case, have omitted any observations that didn't look good.

In fact, once I got that far in my belated analysis of the situation, it occurred to me that it was even possible to make changes in data to have them look better, or to invent data in order to prove a thesis and *then* transfer them to the duplicate pages.

Suddenly, I realized why Professor Dawson had thought that my handing him the shotgun curve was a proof of integrity, and I felt terribly embarrassed.

I like to believe that I have integrity, but that shotgun curve was no proof of it. If it proved anything, it proved only my lack of sophistication.

I felt embarrassed for another reason. I felt embarrassed over having thought it out. For all those years since the shotgun curve, scientific hanky-panky had been literally inconceivable to me, and now I had conceived it, and I felt a little dirty that I had. In fact, I was at this point in the process of changing my career over into full-time writing, and I felt relieved that this was happening. Having now thought of hanky-panky, could I ever trust myself again?

I tried to exorcise the feeling by writing my first straight mystery novel, one in which a research student tampers with his experimental data and is murdered as a direct result. It appeared as an original paperback entitled *The Death-Dealers* (Avon, 1958) and was eventually republished in hardcover under my own title of *A Whiff of Death* (Walker, 1967).

And lately, the subject has been brought to my attention again . . .

Science itself, in the abstract, is a self-correcting, truth-seeking device. There can be mistakes and misconceptions due to incomplete or erroneous data, but the movement is always from the less true to the more true.[1]

Scientists are, however, not science. However glorious, noble and supernaturally incorruptible science is, scientists are, alas, all human.

While it is impolite to suppose that a scientist may be dishonest, and heart-sickening to find out, every once in a while, that one of them is, it is nevertheless something that has to be taken into account.

No scientific observation is really allowed to enter the account books of science until it has been independently confirmed. The reason is that every observer and every instrument has built-in imperfections and biases so that, even assuming perfect integrity, the observation may be flawed. If another observer, with another instrument, and with other imperfections and biases, makes the same observation, then that observation has a reasonable chance of possessing objective truth.

This requirement for independent confirmation also serves, however, to take into account the fact that the assumption of perfect integrity may not hold. It helps us counteract the possibility of scientific dishonesty.

Scientific dishonesty comes in varying degrees of venality; some almost forgivable.

In ancient times, one variety of intellectual dishonesty was that of pretending that what you had produced was actually the product of a notable of the past.

One can see the reason for this. Where books could be produced and multiplied only by painstaking hand copying, not every piece of writing could be handled. Perhaps the only way of presenting your work to the public would be to pretend it had been written by Moses, or Aristotle, or Hippocrates.

---

[1] Lest someone ask me "What is truth?" I will define the measure of "truth" as the extent to which a conception, theory or natural law fits the observed phenomena of the universe.

If the pretender's work is useless and silly, claiming it as the product of a great man of the past confuses scholarship and mangles history until such time as the matter is straightened out.

Particularly tragic, though, is the case of an author who produces a great work for which he forever loses the credit.

Thus, one of the great alchemists was an Arab named Abu Musa Jabir ibn Hayyan (721–815). When his works were translated into Latin, his name was transliterated into Geber and it is in that fashion he is usually spoken of.

Geber, among other things, prepared white lead, acetic acid, ammonium chloride and weak nitric acid. Most important of all, he described his procedures with great care and set the fashion (not always followed) of making it possible for others to repeat his work and see for themselves that his observations were valid.

About 1300, another alchemist lived who made the most important of all alchemical discoveries. He was the first to describe the preparation of sulfuric acid, the most important single industrial chemical used today that is not found as such in nature.

This new alchemist, in order to get himself published, attributed his finding to Geber and it was published under that name. The result? We can speak only of the False Geber. The man who made this great discovery is unknown to us by name, by nationality, even by sex, for the discoverer might conceivably have been a woman.

Much worse is the opposite sin of taking credit for what is not yours.

The classic case involved the victimization of Niccolo Tartaglia (1500–57), an Italian mathematician who was the first to work out a general method for solving cubic equations. In those days, mathematicians posed problems to each other, and upon their ability to solve these problems rested their reputations. Tartaglia could solve problems involving cubic equations and could pose problems of that sort which others found insoluble. It was natural in those days to keep such discoveries secret.

Another Italian mathematician, Girolamo Cardano (1501–76) wheedled the method from Tartaglia under a solemn promise of secrecy—and then published it. Cardano did admit he got it from Tartaglia, but not very loudly, and the method for solving cubic equations is still called Cardano's rule to this day.

In a way, Cardano (who was a great mathematician in his own right) was justified. Scientific findings that are known, but not published, are useless to science as a whole. It is the publishing that is now considered crucial and the credit goes, by general consent, to the first who publishes and not to the first who discovers.

The rule did not exist in Cardano's time, but reading it back in time, Cardano should get the credit anyway.

(Naturally, where publication is delayed through no fault of the discoverer, there can be a tragic loss of credit, and there have been a number of such cases in the history of science. That, however, is an unavoidable side effect of a rule that is, in general, a good one.)

You can justify Cardano's publication a lot easier than his having broken his promise. In other words, scientists might not actually do anything scientifically dishonest and yet behave in an underhanded way in matters involving science.

The English zoologist Richard Owen was, for instance, very much against the Darwinian theory of evolution, largely because Darwin postulated random changes that seemed to deny the existence of purpose in the universe.

To disagree with Darwin was Owen's right. To argue against Darwinian theory in speech and in writing was also his right. It is sleazy, however, to write on the subject in a number of anonymous articles and in those articles quote your own work with reverence and approval.

It is always impressive, of course, to cite authorities. It is far less impressive to cite yourself. To appear to do the former when you are really doing the latter is dishonest—even if you yourself are an accepted authority. There's a psychological difference.

Owen also fed rabble-rousers anti-Darwinian arguments

and sent them into the fray to make emotional or scurrilous points that he would have been ashamed to make himself.

Another type of flaw arises out of the fact that scientists are quite likely to fall in love with their own ideas. It is always an emotional wrench to have to admit one is wrong. One generally writhes, twists and turns in an effort to save one's theory, and hangs on to it long after everyone else has given it up.

That is so human one need scarcely comment on it, but it becomes particularly important to science if the scientist in question has become old, famous and honored.

The prize example is that of the Swede Jöns Jakob Berzelius (1779–1848), one of the greatest chemists in history, who, in his later years, became a powerful force of scientific conservatism. He had worked up a theory of organic structure from which he would not budge, and from which the rest of the chemical world dared not deviate for fear of his thunders.

The French chemist Auguste Laurent (1807–53), in 1836, presented an alternate theory we now know to be nearer the truth. Laurent accumulated firm evidence in favor of his theory and the French chemist Jean Baptiste Dumas (1800–84) was among those who backed him.

Berzelius counterattacked furiously and, not daring to place himself in opposition to the great man, Dumas weaseled out of his former support. Laurent, however, held firm and continued to accumulate evidence. For this he was rewarded by being barred from the more famous laboratories. He is supposed to have contracted tuberculosis as a result of working in poorly heated provincial laboratories and therefore died in middle age.

After Berzelius died, Laurent's theories began to come into fashion and Dumas, recalling his own early backing of them, now tried to claim more than his fair share of the credit, proving himself rather dishonest after having proved himself rather a coward.

The scientific establishment is so often hard to convince

of the value of new ideas that the German physicist Max Planck (1858–1947) once grumbled that the only way to get revolutionary advances in science accepted was to wait for all the old scientists to die.

Then, too, there is such a thing as overeagerness to make some discovery. Even the most staunchly honest scientist may be tempted.

Take the case of diamond. Both graphite and diamond are forms of pure carbon. If graphite is compressed very intensely, its atoms will transform into the diamond configuration. The pressure need not be quite so high if the temperature is raised so that the atoms can move and slip around more easily. How, then, to get the proper combination of high pressure and high temperature?

The French chemist Ferdinand Fréderic Moissan (1852–1907) undertook the task. It occurred to him that carbon would dissolve to some extent in liquid iron. If the molten iron (at a rather high temperature, of course) were allowed to solidify, it would contract as it did so. The contracting iron might exert a high pressure on the dissolved carbon and the combination of high temperature and high pressure might do the trick. If the iron were dissolved away, small diamonds might be found in the residue.

We now understand in detail the conditions under which graphite will change to carbon and we know, beyond doubt, that the conditions of Moissan's experiments were insufficient for the purpose. He could not possibly have produced diamonds.

Except that he did.

In 1893, he exhibited several tiny impure diamonds and a sliver of colorless diamond, over half a millimeter in length, which he said he had manufactured out of graphite.

How was that possible? Could Moissan have been lying? Of what value would that have been to him, since no one could possibly have confirmed the experiment and he himself would know he had lied?

Even so, he might have gone slightly mad on the subject, but most science historians prefer to guess that one of Moissan's assistants introduced the diamonds as a practical joke on the boss. Moissan fell for it, announced it, and the joker could not then back out.

More peculiar still is the case of the French physicist René Prosper Blondlot (1849–1930).

In 1895, the German physicist Wilhelm Konrad Roentgen (1845–1923) had discovered X rays and had, in 1901, received the first Nobel Prize in physics. Other strange radiations had been discovered in that period: cathode rays, canal rays, radioactive rays. Such discoveries led on to scientific glory and Blondlot craved some—which is natural enough.

In 1903, he announced the existence of "N rays" (which he named in this fashion in honor of the University of Nancy, where he worked). He produced them by placing solids such as hardened steel under strain. The rays could be detected and studied by the fact (Blondlot said) that they brightened a screen of phosphorescent paint, which was already faintly luminous. Blondlot claimed he could see the brightening, and some others said they could see it, too.

The major problem was that photographs didn't show the brightening and that no instrument more objective than the eager human eye upheld the claims of brightening. One day, an onlooker privately pocketed an indispensable part of the instrument Blondlot was using. Blondlot, unaware of this, continued to see the brightening and to "demonstrate" his phenomenon. Finally, the onlooker produced the part and a furious Blondlot attempted to strike him.

Was Blondlot a conscious faker? Somehow I think he was not. He merely wanted to believe something desperately—and he did.

Overeagerness to discover or prove something may actually lead to tampering with the data.

Consider the Austrian botanist Gregor Mendel (1822–

84), for instance. He founded the science of genetics and worked out, quite correctly, the basic laws of heredity. He did this by crossing strains of green-pea plants and counting the offspring with various characteristics. He thus discovered, for instance, the three-to-one ratio in the third generation of the cross of a dominant characteristic with a recessive one.

The numbers he got, in the light of later knowledge, seem to be a little too good, however. There should have been more scattering. Some people think, therefore, that he found excuses for correcting the values that deviated too widely from what he found the general rules to be.

That didn't affect the importance of his discoveries, but the subject matter of heredity comes close to the heart of human beings. We are a lot more interested in the relationship between our ancestors and ourselves than we are in diamonds, invisible radiations and the structure of organic compounds.

Thus, some people are anxious to give heredity a major portion of the credit for the characteristics of individual people and of groups of people; while others are anxious to give that credit to the environment. In general, aristocrats and conservatives lean toward heredity; democrats and radicals lean toward environment.[2]

Here one's emotions are very likely to be greatly engaged —to the point of believing that one or the other point of view *ought* to be so whether it is so or not. It apparently takes distressingly little, once you begin to think like that, to lean against the data a little bit.

Suppose one is extremely environmental (far more than I myself am). Heredity becomes a mere trifle. Whatever you inherit you can change through environmental influence and pass on to your children, who may again change them

[2] Since I never pretend to godlike objectivity myself, I tell you right now that I myself lean toward environment.

and so on. This notion of extreme plasticity of organisms is referred to as "the inheritance of acquired characteristics."

The Austrian biologist Paul Kammerer (1880–1926) believed in the inheritance of acquired characteristics. Working with salamanders and toads from 1918 onward, he tried to demonstrate this. For instance, there are some species of toads in which the male has darkly colored thumb-pads. The midwife toad doesn't, but Kammerer attempted to introduce environmental conditions that would cause the male midwife toad to develop those dark thumb-pads even though it had not inherited them.

He claimed to have produced such midwife toads and described them in his papers but would not allow them to be examined closely by other scientists. Some of the midwife toads were finally obtained by scientists, however, and the thumb-pads proved to have been darkened with India ink. Presumably, Kammerer had been driven to do this through the extremity of his desire to "prove" his case. After the exposure, he killed himself.

There are equally strong drives to prove the reverse—to prove that one's intelligence, for instance, is set through heredity and that little can be done in the way of education and civilized treatment to brighten a dumbbell.

This would tend to establish social stability to the benefit of those in the upper rungs of the economic and social ladder. It gives the upper classes the comfortable feeling that those of their fellow humans who are in the mud are there because of their own inherited failings and little need be done for them.

One psychologist who was very influential in this sort of view was Cyril Lodowic Burt (1883–1971). English upper class, educated at Oxford, teaching at both Oxford and Cambridge, he studied the IQs of children and correlated those IQs with the occupational status of the parents: higher professional, lower professional, clerical, skilled labor, semiskilled labor, unskilled labor.

He found that the IQs fit those occupations perfectly. The lower the parent was in the social scale, the lower the IQ of the child. It seemed a perfect demonstration that people should know their place. Since Isaac Asimov was the son of a shopkeeper, Isaac Asimov should expect (on the average) to be a shopkeeper himself, and shouldn't aspire to compete with his betters.

After Burt's death, however, doubts arose concerning his data. There were distinctly suspicious perfections about his statistics.

The suspicions grew and grew and in the September 29, 1978, issue of *Science,* an article appeared entitled, "The Cyril Burt Question: New Findings" by D. D. Dorfman, a professor of psychology at the University of Iowa. The blurb of the article reads: "The eminent Briton is shown, beyond reasonable doubt, to have fabricated data on IQ and social class."

And that's it. Burt, like Kammerer, wanted to believe something, so he invented the data to prove it. At least that's what Professor Dorfman concludes.

Long before I had any suspicions of wrongdoing in connection with Burt, I had written an essay called "Thinking About Thinking" (see *The Planet That Wasn't,* Doubleday, 1976) in which I denounced IQ tests, and expressed my disapproval of those psychologists who thought IQ tests were good enough to determine such things as racial inferiority.

A British psychologist in the forefront of this IQ research was shown the essay by his son, and he was furious. On September 25, 1978, he wrote me a letter in which he insisted that IQ tests were culturally fair and that blacks fall twelve points below whites even when environments and educational opportunities are similar. He suggested I stick to things I knew about.

By the time I got the letter I had seen Dorfman's article in *Science* and noted that the psychologist who had written to me had strongly defended Burt against "McCarthyite character assassination." He also had apparently described Burt as "a deadly critic of other people's work when this departed in any way from the highest standards of accuracy and logical consistency" and that "he could tear to ribbons anything shoddy or inconsistent." It would appear, in other words, that not only was Burt dishonest, but he was a hypocrite in the very area of his dishonesty. (That's not an uncommon situation, I think.)

So, in my brief reply to X, I asked him how much of his work was based on the findings of Cyril Burt.

He wrote me a second letter on October 11. I expected another spirited defense of Burt, but apparently he had grown more cautious concerning him. He told me the question of Burt's work was irrelevant; that he had reanalyzed all the available data, leaving out entirely Burt's contribution; and that it made no difference to the final conclusion.

In my answer I explained that in my opinion Burt's work was totally relevant. It demonstrated that in the field of heredity versus environment, scientists' emotions could be so fiercely engaged that it was possible for one of them to stoop to falsifying results to prove a point.

Clearly, under such conditions, *any* self-serving results must be taken with a grain of salt.

I'm sure that my correspondent is an honest man and I would not for the world cast any doubts upon his work. However, the whole field of human intelligence and its measurement is as yet a gray area. There is so much uncertainty in it that it is quite possible to be full of honesty and integrity and yet come up with results of questionable value.

I simply don't think it is reasonable to use IQ tests to produce results of questionable value which may then serve to justify racists in their own minds and to help bring about the kind of tragedies we have already witnessed earlier in this century.

Clearly, my own views are also suspect. I may well be as
anxious to prove what I want to prove as ever Burt was, but
if I must run the (honest) chance of erring, then I would
rather do so in opposition to racism.

And that's that.

THE PEOPLE

15.

The Unsecret Weapon

Recently, at a rather large meeting of a group of fine people whom I was going to address, I was introduced to others on the dais. On such occasions there is only a certain number of stereotypical remarks one can encounter and I amuse myself at times by responding in a nonstereotypical fashion (if I can think of one).

On this occasion, one of the gentlemen to whom I was introduced held out his hand eagerly and said, "I have heard so much about you."

"Oh, well," I said, modestly, "the ladies *will* talk!"

The gentleman burst into loud laughter and said, "What a great one-liner! Why don't I think of things like that?"

"Why do you have to?" I said. "Use the one I just made up."

"It would be a little difficult," he said. "I'm a Baptist minister."

Just the same, even when they turn out to be a little inappropriate, I love one-liners. I've even got some made up and waiting for questions that will probably never be asked me.

Consider, for instance, the prehistoric days of science fiction and the great part that "secret weapons" played. When jut-jawed Kimball Seaton invents, on Sunday, a planetary pressor that can knock stars to one side without any recoil, builds it on Monday and uses it on Tuesday, that's enough (a) to ruin the vicious reptilian Sandivorians and (b) to ravish the soul of the reader with delight.

But you know, science fiction doesn't invent things out of nothing, usually. There is some hint of even the wildest concoctions in real life, and there have indeed been secret weapons in actual history.

So there you are: I am waiting for someone to ask me, "Dr. Asimov, what was the most remarkable secret weapon in history?"

And my cute one-liner will be, "One that wasn't secret."

Let me explain. Any weapon can be secret if the enemy happens not to know about it till it is used.

If the two combatants are on a technological par, however, the mere fact that the weapon is used gives it away and in a surprisingly short time the enemy has it, too.

Thus, in World War I, the Germans used poison gas as a secret weapon and the Allies used tanks. In both cases, the first attack making use of the secret weapon was effective but before long the other side had it, too.

Even when the secret weapon is extremely complicated and extremely unprecedented and the details of its structure have been kept extremely secret, it can be duplicated with surprising speed. In 1945, the Americans used the nuclear fission bomb on the Japanese—and by 1949, the Soviet Union had it, too.

In order to confine our discussion to true secret weapons,

then, we ought to look for those that are not duplicated by the enemy for a considerable period of time, even after it is used and its existence revealed.

And, mind you, we're talking about combatants who are in a state of reasonable technological equivalence. Gunpowder weapons were, effectively, secret weapons to the Indians when Europeans arrived on the American continents. Though Indians learned to use guns, they never learned to make guns for themselves—so Europeans and their descendants took over two continents.

If we stick to weapons that are secret even after being used, and that technologically equivalent enemies do not adopt even though they are being defeated by them, then there is one, and only one that I can think of, that is truly secret. It was used by a single nation on a number of different occasions spread over a substantial period of time and was never duplicated by any other nation. In fact, it remains secret *to this day*. It's "Greek fire."

We *guess* that Greek fire was some combination of sulfur, naphtha, quicklime (calcium oxide) and niter (potassium nitrate). Naphtha is a hydrocarbon mixture found naturally in the Middle East that is not too different from modern gasoline.

When water is added to Greek fire, it reacts with the calcium oxide and develops considerable heat in the process—enough heat to ignite the naphtha in the presence of oxygen released from the potassium nitrate. This, in turn, ignites the sulfur, making it burn and produce choking vapors of sulfur dioxide.

If the Greek-fire mixture is placed in brass-bound wooden tubes and if a jet of water hits it from behind, it will burst into flame. The push of the water and the expansion of the exhaust gases formed will combine to fling the burning mixture out of the tube for considerable distances. If the burning mixture hits the ocean surface, it will float and it will burn all the more fiercely.

Imagine, then, that a seaport is being attacked by an

enemy fleet at a time when all ships are made of wood. If you are on one of the ships of the enemy fleet, you will see flame jet toward you emitting choking fumes. What is really horrifying is that it is not extinguished by water, but continues to float toward you so that it will eventually set fire to the ship at the waterline.

The terror of the weapon itself will demoralize the attackers, multiplying the effect of what actual burning of the ships there is.

The inventor of Greek fire is supposed to have been one Callinicus, concerning whom, aside from the invention, precisely nothing is known, not even whether he was born in Syria or Egypt. Apparently, he was born in one of those provinces and, when they fell to the Arabs about A.D. 640, he fled to Constantinople and there, in the fullness of time, produced the mixture.

By 669, the triumphant Arabs, all aglow with the brand-new faith of Islam, had overrun Asia Minor and were just across the narrow strait from Constantinople. The Byzantine Empire, of which Constantinople was the capital, was staggering under multiple catastrophes and all that kept the city safe was the Byzantine fleet.

But the Arabs had learned how to build and handle ships, too, and in 672, an Arab fleet approached the great city. If the Arab fleet could overcome the sea defenses of Constantinople, the city would fall and, with it, what was left of the empire. The Arabs sweeping through the Balkans would find nothing in the moribund Europe of the Dark Ages to stop them. Just as Iran, Iraq, Syria and Egypt were being permanently converted to Islam, so would Europe have been.

Except that Constantinople had Greek fire. In 672 it was used for the first time, the Arab ships burned, the Arab seamen panicked and Constantinople was saved. And to those who felt it important that Europe remain Christian, this was a heaven-sent miracle.

When the Arabs returned to the assault in 717, their ships were again repelled by Greek fire, and Constantinople was again saved.

Greek fire was used on occasion in other naval engagements over the following century and then, for some reason, went out of use with its secret still inviolate.

One can understand the reason why Greek fire was secret. It was a complicated chemical mixture that others saw only as it was burning. Without an unburned sample to study and with chemical technology still in an embryonic stage, it is not surprising that no one could duplicate it, or even dream of duplicating it.

But I have another weapon in mind that was every bit as terrifying and effective as Greek fire and yet was so simple that anyone could see what it was—how to make it, how to use it and everything else about it. It was, therefore, not really secret except that no one (with a single exception I'll come to) copied it and adopted it. They merely confined their reaction to being defeated by it.

Since prehistoric times, the best and most efficient long-range weapon was the bow and arrow. (There was also the sling, but it never attained anything like comparable popularity.)

The bow and arrow was such a simple, straightforward weapon that it was difficult to improve on. About the only thing one could do was to make the wood of the bow stiffer and the bowstring stronger so that when it was deformed and then released, the return to normal would be faster and the arrow would be sent at a greater speed and therefore over a longer distance and with more penetrating power. The difficulty was that the more forcefully the bow sprang back to normal, the harder it was to deform in the first place. (You get nothing for nothing.)

Sometime about 1000, in Italy, a new kind of bow was developed—one that was made of metal and not of wood. It was entirely too stiff to be bent by human muscle. The metal bow was therefore attached to a metal crosspiece (so that the bow as a whole looked like a cross, and was called a crossbow). The crosspiece contained a groove into which a metal arrow (or "bolt") could be placed.

The bowstring was not pulled back by hand, but by a crank attached to the crosspiece. The archer turned the crank until the bowstring was pulled far enough back, fixed it in place, put the bolt in the groove and pushed the release lever, and off went the bolt flying with much greater force than an ordinary arrow would. That bolt had a range of a thousand feet and, at closer distances, could penetrate armor.

It was a very easy weapon to learn to use and could be handled from any position. It was a greatly feared weapon and, in 1139, a church council outlawed its use as being too horrible—at least among Christians. It was decided that it might lawfully be used against non-Christians. (Lest you worry about this example of bigotry, let me assure you the edict was a dead letter. Christian armies, scorning all prejudice, used crossbows freely against other Christian armies.)

The crossbow had the disadvantage, however, of taking a long time to reload. Once it had been fired, it had to be fixed against the ground or in some other firm position, slowly cranked up to the necessary pitch and the bolt fitted. While that was done, the crossbowman was vulnerable to enemy attack. (We still speak of someone who has used up his talent or wit or ability as having "shot his bolt.")

I am not, however, thinking of the crossbow as a secret weapon. It was quickly adopted by other nations, who either trained their own corps or hired Italian mercenaries.

The secret weapon was another variety of the bow and arrow; one that remained of wood, but increased its size and stiffness until its use required the limit of human strength. It was the "longbow," so called because it was six feet long or more and shot arrows that were a yard long, the famous "cloth yard shafts."[1]

The longbow was lighter than the crossbow and had an even longer range, up to twelve hundred feet at maximum. Much more important, the longbow could be fired very rapidly. The longbowman, reaching over his shoulder for arrows in the quiver he carried on his back, could fire five or six accurate shots in the time it took the crossbowman to reload.

The result was that if equal numbers of longbowmen and crossbowmen encountered each other, the latter were sure to be riddled.

In fact, the longbow was the most deadly and versatile weapon that was to be seen in war until such time as gunpowder weapons became efficient. Several thousand longbowmen shooting at once could produce a cloud of death, dropping from the sky with a hissing sound, that simply could not be withstood.

If the longbow had a tactical disadvantage, it was that it was a long-range weapon. If the enemy could get close enough to the longbowmen, the latter could be chopped up. The trick was to get close enough and still be alive, something that was never managed without gunpowder weapons.

Yet how could the longbow remain effective? Anyone could see what it was. Anyone could duplicate it. To be sure, the best wood for the longbow was English yew, which didn't grow everywhere, but I daresay other kinds of wood could have been used and found to be good enough.

What, then, made this unsecret weapon effectively secret? What kept nations who were defeated by the longbow from adopting the weapon?

---

[1] In ballads, stories and movies, Robin Hood and his merry men are invariably shown as using the longbow, which was unknown in the time of Bad King John. Sorry!

Two things. First, there was the small matter of training. The longbow was stiff. It took a pull of very nearly a hundred pounds to stretch it. It took years of training and a strong pair of arms and shoulders to pull the string back to the ear with one smooth motion, so that the arrow could be loosed with greater force than the bolt of a crossbow, and only one nation was willing to invest in the training.

Second, the crossbow could be handled by anyone, so that crossbowmen were easily and quickly trained and, since they were lowborn rabble, could be treated like lowborn rabble. They could always be replaced.

Longbowmen, however, though equally lowborn, were the product of years of training and could not be easily replaced. They had to be cherished and conserved and treated like so many jewels.

A particularly aristocratic army, therefore, would find it psychologically difficult to develop a longbow corps. They would rather lose a battle in knightly fashion than owe a victory to rabble.

The longbow was invented in Wales at some unknown time and by some unknown Welshman. The Welsh, stubborn fighters, had held off first the Saxons and then the Normans, ever since King Arthur's time, but in 1272, Edward I came to the English throne. He was the most capable crowned warrior since William the Conqueror and it was his intention to mop up the Welsh.

In 1282, he began a two-year campaign in Wales and encountered the longbow in enemy hands. Fortunately for him, the Welsh were relatively few in number and did not use the weapon *en masse* and with discipline. Edward won the war, adopted the longbow and set about training a large corps of men who would use that weapon properly. (That was the first and last time that an army adopted the longbow after encountering it in the hands of an enemy. I have never seen Edward I given due credit for this.)

Once the Welsh were conquered, Edward I turned to

Scotland, which was in anarchy. After he had reduced it to a puppet kingdom, the Scots rose in rebellion under William Wallace and, on July 22, 1298, Edward I met Wallace's army at the Battle of Falkirk.

The Scots were hardy and brave fighters and faced Edward with twenty-five thousand pikemen, whose long heavy pikes, or spears, converted them into a formidable and massive porcupine. The English cavalry drove off the less numerous Scottish cavalry but could make no dent in the pikes.

Edward I then unleashed his new weapon for the first time. His longbowmen, from a distance, loosed their volleys, and the Scottish pikemen crumbled. They could not fight back against the distant enemy and they died in droves. The English cavalry charged again and the Scots were wiped out.

For a while, it looked as though Scotland, like Wales, would pass under English dominion. Under Robert Bruce, however, Scotland rebelled again. Grim Edward I was marching north to teach the stubborn Scots another lesson in 1307 but died en route. His son, the unwarlike Edward II, called off the invasion.

The pressure of events, however, and English public opinion forced Edward II to invade Scotland seven years later and, at Bannockburn, on June 24, 1314, he met the forces of Robert Bruce. Between Bruce's clever maneuvering and Edward's stupid handling of his own army, the English ended with the longbowmen crowded behind their own cavalry.

The English cavalry could make no impact on the Scottish pikemen, and the longbowmen could not get a clear shot at the enemy. When they tried to fire in high arcs over their own cavalry, the maneuver failed and it was the cavalry that suffered.

In the end it was a smashing Scottish victory and Scottish independence was saved. Between 1298 and 1547—two and a half centuries—there were many battles fought between

the Scots and the English, and the English won every single
one, except Bannockburn. That one loss was enough.

The true triumph of the longbow came in France, however.
For reasons it would be tedious to go into here, Edward II's
son, Edward III, had a very good claim to the French
throne. There was only one serious flaw in the genealogical
argument and that was that the French people didn't want
an English king, but in those days that was considered irrelevant.

In 1337, Edward III declared war on France and in 1340,
he won an important naval victory and gained control of
the English Channel. It wasn't till 1346, however, that he
could actually scrape together the men and money with
which to invade France. He intended only a demonstration,
but when he tried to make his way back to England, the
French army, in pursuit, caught up to him at Crécy, a town
near Calais, where the English Channel was narrowest.

The French king, Philip VI, had about 60,000 men, which
included 12,000 armored knights and 6000 skilled Genoese
crossbowmen.

Edward III had only about 12,000 men, but these included 8000 well-trained longbowmen. The longbowmen
were carefully distributed along the line of battle, with 4000
knights relegated to the minor role of protecting them. Pitfalls were dug before the line of longbowmen to serve as
further protection in case the enemy got that far.

As soon as the French army arrived, the knights clamored
to charge the lowborn English rabble who were so few in
number, even though it was already late in the day and it
would have made more sense to get a night's rest first. The
Genoese crossbowmen pointed that out and explained they
had just finished an exhausting march. The knights, however (who were on horseback), called the crossbowmen
cowards and ordered them forward.

The crossbowmen advanced against an English army that
had been carefully arranged to have the afternoon sun

behind them and full in the eyes of the advancing Genoese. The cloth yard shafts converged on the crossbowmen before they could advance within the range of their own weapons and they had no choice but to retreat hastily.

This enraged the French knights, who pushed forward in a ragged line, even though no order to charge had been given. The cry rang out, "Run those cowardly rascals down; they but impede progress." The cavalry thereupon trampled over their own crossbowmen and spurred their horses toward the English.

The English found themselves facing not an army but a mob. It was a brave mob, for the French launched some sixteen charges, but bravery didn't help them. The longbowmen shot volley after volley, and the knights went down in heaps. Before the sun had set, 1550 French knights were dead on the field, while English casualties were insignificant.

If the French thought Crécy was an accident, they were disabused ten years later when, under Philip VI's son and successor, John II, a French army attacked an English army under Edward III's son, the so-called Black Prince, at Poitiers on September 19, 1356.

The battle went precisely as the earlier one had. The outnumbered English used their longbowmen to mow the French knights down.

There followed a long pause. Edward III, after a feckless old age, and the Black Prince, too, both died in 1377. The Black Prince's young son succeeded as Richard II and was finally overthrown by his cousin, who ruled as Henry IV and who had to face civil wars of his own.

Meanwhile, the French, who no longer dared meet the English in the field, had taken up a kind of guerrilla action under a brilliant leader, Bertrand du Guesclin, and retook much of the English conquests. The French never tried to duplicate the longbow, however, even though du Guesclin chanced a pitched battle against the English across the border in Spain and was defeated.

It wasn't till the reign of Henry IV's son, Henry V, that England could turn its full attention to France once more.

On August 14, 1415, Henry V landed a force of 30,000 men at Harfleur, France's chief port in Normandy, and 24,000 of those men were longbowmen. Longbowmen weren't of much use in knocking down city walls, however, and Henry had brought cannon for that purpose. (In fact, Edward III had used very primitive cannon at Crécy.)

Cannon were still rather bumbling weapons, more dangerous to the gunners who fired them than to the enemy, so that it took five weeks to reduce the city; weeks during which Henry's forces were much weakened through attrition and disease.

Once Harfleur was taken, Henry V determined to make his way overland to Calais, which Edward III had taken after the Battle of Crécy, and which was now England's chief stronghold in France. There, Henry intended to allow his men to rest and recuperate while he gathered reinforcement from England.

The march to Calais, however, was a hard one. It rained constantly, the English army continued to dwindle, and to suffer badly from dysentery.

The French followed the English army, waiting for it to weaken sufficiently, and finally trapped it at Agincourt, about thirty-five miles south of Calais (and only twenty miles northeast of Crécy). By this time, the English were reduced to a pitiful 9000 men, bedraggled and sick, while facing them were over 30,000 Frenchmen. The date was October 25, 1415. Sixty years had passed since Poitiers and the Frenchmen were confident again.

Henry was a good general. He chose the site of battle carefully, drawing his thin line of men across a front no more than a thousand yards wide, with either flank blocked off by dense woods. The French would be forced to crowd their men together to attack and would be sure to get in each other's way.

What's more, of Henry's small army, almost all were longbowmen, and they waited for their prey, with the pitfalls in front of them and with sharpened sticks buried in the soil, points upward, to greet any horses that might arrive.

Henry noted, too, that the constant rains, which had caused his army so much suffering, had turned the field into a quagmire. He didn't think the heavily armored knights, either on foot or on horseback, would be able to advance easily.

Of course, if the French chose to wait, the English would be forced to surrender, or to leave their lines to face destruction. The French, however, would not wait in the face of such a tiny army (as Henry knew they would not).

Agincourt is treated as a near-miracle in some accounts, but it wasn't. The French didn't have a chance; it would have been a miracle if the English had lost.

The French charged—or tried to—and were instantly mired in the mud. They were in utter disorder and once they had managed to work their way within range, Henry gave the signal and 8000 cloth yard shafts hissed their way toward the enemy and landed in their crowded ranks. It was impossible to miss, and according to the jubilant accounts of the English, 10,000 Frenchmen died against thirteen English. Even allowing for exaggeration, however, it was an immensely one-sided victory.

Henry V went on, a couple of years later, to capture Normandy and Paris. He forced the French king, Charles VI, to recognize Henry V as his successor.

Henry V died in 1422, however, at the age of thirty-five and there was no Englishman who could handle armies quite as well as he could. Still, the French lost one more large battle to the English longbow at Verneuil on August 17, 1424.

The English placed Orléans under siege in 1428 and it seemed they had only to take that town to force complete domination over a thoroughly demoralized France. England had, however, reached the limit of its strength by now and could not manage to close the siege lines around the city.

French soldiers managed to slip into the city and soon it
was only the superstitious fear and dread of the English and
their longbows that kept the French from breaking out.

It was at this point that Joan of Arc appeared on the
scene and supplied the inspiration necessary for the French
to drive the English away from Orléans. For the over-
strained English, their awe of the "witch" was the final
straw.

The war continued for an additional quarter century,
however, and what decided it at last was something that
finally overtopped the longbow. Charles VII, the new king
of France, supported two brothers, Jean and Gaspard Bu-
reau, who improved the design of cannon and bettered the
quality of gunpowder.

Charles began to build an elaborate artillery arm, the first
effective one in history. Gunners were trained to handle the
cannon and (most important of all) the French knights
were forced to treat the gunners, who were, after all, as lowborn as archers were, with respect. From this point on, it
was the artillery that decided the battles and the reign of
the longbow was over.

The English were as unable to adjust their thinking to the
new artillery as the French had earlier been unable to adjust their thinking to the longbow. By 1453, the English
were driven out of France (all except Calais, which they
held for another century). They never did figure out why
the victories had ceased, either; the general English theory
was that they lost France through a combination of treason
at home and witchcraft in France. (Shakespeare's *Henry
VI, Part One* expresses that view perfectly—a century and a
half after the end of the war.)

This does not mean the total American population is dropping. It is still rising, although at a slower rate than that of the world in general. In 1967, the population of the United States was about 197,600,000, and in 1979 it was about 218,000,000, a gain of just over 10 per cent.

The percentage of Americans living in the great cities declined from 9.65 per cent in 1967 to 8.17 per cent in 1979, but this does not mean that the United States is growing less urbanized or more rural.

The population that is leaving the great cities (and the large cities generally) is flooding into the suburbs at the rim of the city, which are part of the "metropolitan area" marked off from the central city by arbitrary political lines intended to give the suburbs the benefit of the city without responsibility for its problems.

The metropolitan areas are continuing to grow and there are about forty in the United States that have populations of more than a million.

In my 1967 essay, I moved on to the three nations that were more populous than the United States. They were, then, China, India and the Soviet Union in that order, and it remains true today. Here are the comparative statistics of the four top nations:

Table 2. Population of the Most Populous Nations

WORLD RANKING	NATION	POPULATION		% INCREASE
		1979	1967	
1	China	973,334,000	750,000,000	29.8
2	India	649,354,000	475,000,000	36.7
3	Soviet Union	260,178,000	230,000,000	13.1
4	United States	217,799,000	197,600,000	10.2

16.
More Crowded!

I don't often make specific long-range promises in these essays. Sometimes I say, specifically, that I will discuss a particular subject further in the next essay—very short-range. Sometimes I say that something is a subject to be taken up "another time"—very unspecific.

In my essay "Crowded!" (see *Science, Numbers and I,* Doubleday, 1968), however, I discussed some aspects of the population problem as related to the large cities of the world and concluded the article with the following paragraph:

"Where it will all end, I don't know. I can only wait in terror as each day is more crowded than the one before. Ten years from now—if we are all still alive—I shall return to this theme and see how things have progressed."

Well, it's time, especially since I have just received a new book of statistics, *The Book of World Rankings*, by George Thomas Kurian (Facts on File, 1979).

Mr. Kurian drew upon the best international statistics available (sometimes admittedly imperfect) and I will make use, with gratitude and thanks, of his labors. Using it I will see what has happened to us with respect to city population in twelve years.

First, to set the background (and this is not material out of Kurian) . . . From what I have been able to gather, the world population was something like 3.3 billion in 1967 and something like 4.12 billion in 1979. We have increased the world population in the last twelve years by 800 million mouths, or 25 per cent. To put it still another way, we have added another China to the population of the world.

It is quite likely that we will end the decade of the nineteen-eighties with a world population edging toward 5 billion, having added still another China. The population growth of the nineteen-seventies has indeed been terrifying and has contributed enormously to the change for the worse in world economy and social structure in the last twelve years.

The population growth that could now take place in the nineteen-eighties is quite likely to be catastrophic. And having said that, let us go on to the cities.

In 1967, I was living in a Boston suburb and, lacking specific statistics, I guessed that Boston was no better than 151st in the world ranking of cities. I believe that was a pretty good guess but Kurian, in his 313th table, lists all the cities with a population of over 500,000 in the world. There are 287 cities in that list and Boston, with a population given as 636,725 is listed in 207th place.

In my earlier article, I defined a great city as one which contained a population of more than a million, and in 1967, I listed six American cities as great cities. At the present moment, those six are still great cities and no new ones have

been added in the United States. Here are the comparative statistics:

Table 1. The Great Cities of the United States

WORLD RANKING	AMERICAN RANKING	GREAT CITY	POPULATION[1] 1979	1967
5	1	New York	7,481,613	8,080,000
23	2	Chicago	3,099,391	3,520,000
29	3	Los Angeles	2,727,399	2,740,000
50	4	Philadelphia	1,815,808	2,030,000
76	5	Houston	1,357,394	1,100,000
78	6	Detroit	1,337,557	1,600,000

Notice that the population of five out of the six cities has declined in the last twelve years. Detroit is the most extreme, having lost one sixth of its population, and fallen behind Houston, the only American great city to have gained population in the interval.

The phenomenon of urban population loss is common to many American cities. Thus, in 1967, I asked the readers to identify the largest American city that was not a great city. The answer then was Baltimore, and that answer still holds, but Baltimore's population has also declined, from 925,000 to 851,698, a loss of nearly 8 per cent.

The total population of the American great cities was, in 1967, 19,070,000. In 1979, it was 17,800,000 for a decline of nearly 7 per cent.

[1] I don't consider the population statistics in this article to be necessarily accurate to the last digit, or that what I call 1979 represents that year exactly or 1967 that year. Different cities are counted with different degrees of care and accuracy in different years and estimates in the absence of reliable censuses can be wrong, too. On the whole, though, I think the figures in this article represent the correct essence of what is happening to population these last twelve years.

At the present time, China possesses 23.6 per cent of the world's population and India, 15.7 per cent. The four most populous nations of the world contain a total population of just over 2,100,000,000, or almost exactly half the number of people in the world.

Each of the three nations with populations exceeding that of the United States had more great cities in 1967 than we had—and they still do.

In 1967, with what statistics I had at hand I located no fewer than sixteen great cities in China, each with a population of more than a million. In Kurian's book I can find only fourteen great cities listed.[2] I suspect this reflects a general improvement in the accuracy of Chinese statistics available to the rest of the world in the last dozen years.

There is one impossibility in Kurian's table, however. In 1967, I listed the city of Shanghai as the largest of the Chinese great cities, with a population of about 7,000,000, and the latest figures I have, other than Kurian's list, gives it a population of over 10,000,000 now. In Kurian's list, however, Shanghai is placed in 97th place (!) among the cities of the world, with a population of 1,082,000. I can only assume that a zero dropped out of the figure and that the computer that prepared the list followed its instructions and placed Shanghai impossibly far down and that no human proofreader noticed. I feel the population should be 10,082,000 and that is the figure I am going to use, changing the "world ranking" figures that Kurian gives accordingly.

I will save space by listing only those Chinese great cities with populations of over 2,000,000. There are six of these compared with four for the United States.

---

[2] It actually contains fifteen, but it lists "Nagoya, China," which is a clear misprint for "Nagoya, Japan," and I made the correction in my copy.

Table 3. The Most Populous Cities in China

WORLD RANKING	CITY	POPULATION 1979	1967	% CHANGE
1	Shanghai	10,082,000	7,000,000	+44.0
4	Peking	7,570,000	6,800,000	+11.3
13	Tientsin	4,280,000	2,900,000	+47.6
35	Mukden	2,411,000	3,100,000	−22.2
41	Wuhan	2,146,000	(unlisted)	
42	Chungking	2,121,000	2,200,000	− 3.6

Harbon and Canton, listed as over 2,000,000 in 1967, are listed by Kurian as under that figure now. I suspect that my 1967 figures were not necessarily very accurate in connection with the Chinese cities.

The total population of the fourteen great cities of China in Kurian's list is 39,500,000 as compared with 38,000,000 for the sixteen great cities I listed in 1967. The great city population of China is 2.2 times as great as that of the great city population of the United States, but that is not as great as one would expect from the disparity in total population. After all, the total population of China is 4.2 times that of the United States.

Although 8.17 per cent of the American population lives in the great cities, only 4.0 per cent of the Chinese population does.

As for India, Kurian lists eight great cities, as compared to the six I had listed in 1967. All six of my 1967 list are included, and the cities of Bangalore and Kanpur are included in addition. The four largest Indian cities have populations of over 2,000,000 and here they are:

Table 4. The Most Populous Cities in India

WORLD RANKING	CITY	POPULATION		
		1979	1967	% CHANGE
8	Bombay	5,970,575	4,540,000	+31.5
20	Delhi	3,287,883	2,300,000	+43.0
22	Calcutta	3,148,746	3,005,000	+ 4.8
34	Madras	2,469,449	1,840,000	+34.2

Since 1967, Madras has graduated into the 2,000,000 rank and none has fallen out of it. The total population of the eight great cities of India is about 20,750,000 (3.2 per cent of the total population) as against 14,000,000 in 1967 (2.2 per cent of the total population.)

This brings us to the Soviet Union for which, in 1967, I listed seven great cities. Kurian's table, however, lists no fewer than twelve, which puts the Soviet Union second only to China in this respect. In addition to the seven I listed in 1967, we now have Kuibyshev, Sverdlovsk, Tbilisi, Odessa and Omsk.

Only three of the great cities of the Soviet Union, have populations of over 2,000,000 (compared with 2 in 1967). Here they are:

Table 5. The Most Populous Cities in the Soviet Union

WORLD RANKING	CITY	POPULATION		
		1979	1967	% CHANGE
7	Moscow	6,941,961	6,334,000	+ 9.6
18	Leningrad	3,512,974	3,218,000	+ 9.2
45	Kiev	2,103,000	1,292,000	+62.8

The total population of the twelve great cities of the Soviet Union is 23,600,000, as compared with 15,000,000 for the seven great cities of 1967. The percentage of the population living in great cities was 9.0 in 1979 as compared with 6.5 in 1967.

In the world ranking of population, Indonesia is in fifth place immediately behind the United States, and Japan is in sixth place. For the purposes of this article, which deals with cities, we can pass by Indonesia as a nonindustrial nation, and move on to Japan, which is highly industrialized and citified. In fact, we can lump Japan with the four most populous nations and call them the Big Five.

Japan's total population in 1979 is listed in Kurian as 114,595,000 as compared to 96,200,000 in 1967, a rise of 19.1 per cent. In 1967, I listed seven great cities in Japan. Kurian's list shows eight, made up of my seven plus the city of Sapporo. Four of the Japanese great cities now have populations of over 2,000,000, as compared with two in 1967, and here they are:

Table 6. The Most Populous Cities in Japan

WORLD RANKING	CITY	POPULATION 1979	1967	% CHANGE
3	Tokyo	8,442,634	8,730,000	− 3.3
30	Osaka	2,714,642	3,200,000	−15.2
32	Yokohama	2,610,124	1,600,000	+63.1
43	Nagoya	2,083,111	1,900,000	+ 9.6

The total population of the great cities of Japan is about 20,860,000 in 1979, as compared to 18,800,000 in 1967. The percentage of the Japanese population that lives in the great cities is 18.2 in 1979 as compared to 19.5 in 1967. This is a small decline, but the percentage of the great city population is nevertheless higher in Japan than in any of the others of the Big Five.

Suppose we consider the Big Five nations together. The total number of great cities in the Big Five is 48 compared to the 42 I listed in 1967.

In 1967, however, I pointed out that there were 46 great cities remaining in nations other than the Big Five. In Kurian's list there are 61 great cities in those other nations. In other words, there were 88 great cities in the world in 1967 and 109 in 1979, an increase of 23.9 per cent. Dividing them by continents, this is what happens:

Table 7. Great Cities of the World

CONTINENT	GREAT CITIES		TOTAL POPULATION		
	1979	1967	1979	1967	% CHANGE
Asia	52	42	126,900,000	91,700,000	+ 38.4
Europe	28	25	60,900,000	51,300,000	+ 18.7
North America	11	9	31,400,000	24,200,000	+ 29.7
South America	10	7	26,260,000	15,100,000	+ 73.9
Africa	6	3	13,660,000	6,100,000	+124
Australia	2	2	5,520,000	4,300,000	+ 28.3
TOTAL	109	88	264,640,000	192,700,000	+ 37.3

As you see, the great city population is increasing somewhat faster than the general world population, and this is particularly true in South America and in Africa. In 1967, 5.84 per cent of the world population (1 in 17) lived in a great city. In 1979, 6.42 per cent (1 in 15.5) did.

In my 1967 essay, I asked which was the largest nation that did not contain a great city. The answer I then gave was Nigeria, which, I said, had a population of 56,400,000,

while its largest city and capital, Lagos, had a population of
only 665,000.

Well, Nigeria's population is now 67,520,000 and Lagos,
with a population of 1,061,221, is one of the great cities.
The new candidate, if we go by Kurian's tables is South
Africa, with a population of 25,003,000 and with Durban its
largest city at 730,000.

If you look through the great cities I have listed so far in
the Big Five, you will notice that Shanghai, China is in first
place in the World Ranking, while Tokyo is in third place,
Peking in fourth, and New York in fifth.

Second place is missing, so it has to be a city in a nation
that is not one of the Big Five. I wonder how many of you
can guess what the second largest city in the world happens
to be at the moment (at least according to Kurian's tables).
Frankly, I would not have guessed it—and it isn't London, if
any of you have guessed that.

I'll tell you. It's Mexico City. In 1967, I listed its population as 3,193,000, which means an increase of about 170 per
cent. This seems hard to believe and it may be that Mexico
City has, in the interval, absorbed some of its suburbs. Even
so, the fact is, it is growing at a phenomenal rate.

Here is the list of the ten largest cities in the world:

Table 8. The Largest Cities in the World

WORLD RANKING		CITY	POPULATION		
1979	1967		1979	1967	% CHANGE
1	4	Shanghai, China	10,082,000	7,000,000	+ 44.0
2	13	Mexico City, Mexico	8,628,024	3,193,000	+170.2
3	1	Tokyo, Japan	8,442,634	8,730,000	− 3.3

4	5	Peking, China	7,570,000	6,800,000	+ 11.3
5	3	New York, U.S.	7,481,613	8,080,000	− 7.4
6	2	London, U.K.	7,167,600	8,185,000	− 12.5
7	6	Moscow, U.S.S.R.	6,941,961	6,334,000	+ 9.6
8	7	Bombay, India	5,970,575	4,540,000	+ 31.5
9	9	Cairo, Egypt	5,715,000	3,518,000	+ 62.5
10	19	Jakarta, Indonesia	5,476,009	2,907,000	+ 88.4

As you see, there are two newcomers to the list of the big ten in the last dozen years, Mexico City and Jakarta. The two that have dropped out to make room are Chicago, which was in eighth place in 1967, and Leningrad, which was in tenth place in 1967.

China is the only nation which places two cities in the top ten, though in 1967, the Soviet Union and the United States also did. By continents, in 1979, five great cities are Asian, two are European, two are North American and one is African. In 1967, the figures were four, three, two and one respectively.

The total population of the top ten cities is about 73,500,000 in 1979, or about 1.8 per cent of the population of the world. In 1967, it was 59,900,000, also about 1.8 per cent of the population of the world. No change there.

I don't want to leave you without suggesting a parlor game guaranteed to keep your guests (if they are the intellectual parlor game type) busy for an entire evening. If you supply them with some drinks to sip at, and paper and pen, you can then slip away and go to a movie.

It is simple: Just ask them to make two columns. In one column, they can list the largest city in the world that be-

gins with each of the letters of the alphabet in order; in the
second, the largest city in the United States.

I'll give *you* the answers, naturally, and here they are:

A—That's a difficult one right away. It is Alexandria,
Egypt, with a population of 2,259,000. The largest American
city beginning with *A* is Atlanta, Georgia, population
436,000.

B—Bombay, India, 5,970,575. If you want to eliminate
that because it's too easy; second largest is Berlin, if you
count East and West together, making it 4,085,960. If you
don't want to allow the combination, that brings you to
Buenos Aires at 2,972,453. The largest American *B* is Baltimore, Maryland, 851,698.

C—Cairo, Egypt, 5,715,000. The largest American *C* is, as
anyone can guess, Chicago, Illinois, 3,099,391. If Chicago is
too easy, then try the second largest, which is Cleveland,
Ohio, 638,693.

D—Delhi, India, 3,287,883. The largest American *D* is Detroit, Michigan, 1,335,085. If that's too easy, the next largest
is Dallas, Texas, 812,797.

E—This one isn't at all easy. It is Erevan, Soviet Union,
928,000. I've seen it spelled Yerevan and if it is disqualified
for that reason then the next largest is Essen, West Germany, 677,508. The largest American *E* is El Paso, Texas,
385,691.

F—One of the hardest in the list. The largest is Fushun,
China, 985,000. And if you miss that, you won't get the sec-
ond largest either, which is Fukuoka, Japan, 964,755. The
American *F* is Fort Worth, Texas, 358,364.

G—Guadalajara, Mexico, 1,640,902. The American *G* is
Grand Rapids, Michigan, 187,946.

H—Ho Chi Minh City, Vietnam, 1,825,297. If you're one
of those old-fashioned souls who insists on calling it Saigon,
then you'll have to pass on to the next largest, which is
Hamburg, West Germany, 1,717,383. The American *H* is, of

course, Houston, Texas, 1,357,394. If you disqualify that as too easy, then the next in line is Honolulu, Hawaii, 324,871.

I—Istanbul, Turkey, 2,376,296. Some people might want to disqualify it because it is really Constantinople—but it isn't. It's name hasn't been officially Constantinople for five hundred years. The American *I* is Indianapolis, Indiana, 735,077. As a matter of curiosity, the next in line is (guess!) Independence, Missouri, 111,481.

J—Jakarta, Indonesia, 5,476,009. I've seen it spelled Djakarta and if anyone uses it as the largest *D*, then that leaves the next place for Johannesburg, South Africa, 654,682. The American *J* is Jacksonville, Florida, 562,283.

K—Karachi, Pakistan, 3,498,634. The American *K* is Kansas City, Missouri, 472,529.

L—London, of course, 7,167,000. If you want to eliminate that as too easy, you must move down to Leningrad, 3,513,974. The American *L* is Los Angeles, 2,727,399, and if you want to eliminate *that*, then it's Long Beach, California, or Louisville, Kentucky, in a virtual tie at about 336,000.

M—Mexico City, of course, 8,628,024, then Moscow, 6,941,961, and then Madrid, Spain, 3,520,320. The American *M* is Milwaukee, Wisconsin, 665,796, with Memphis, Tennessee right behind at 661,319.

N—Yes, New York, 7,481,613, with Nagoya, Japan, 2,083,111 in second place. The American *N* in second place is New Orleans, Louisiana, 559,770.

O—Osaka, Japan, 2,714,642. The American *O* is Omaha, Nebraska, 371,455.

P—Peking, China, 7,570,000, with Paris in second place, 2,290,000. The American *P* is Philadelphia, Pennsylvania, 1,815,808, and in second place is Phoenix, Arizona, 664,721.

Q—Quezon City, Philippines, 994,679. The American *Q* is Quincy, Massachusetts, 91,494.

R—Rio de Janeiro, Brazil, 4,252,009. In second place is Rome, 2,868,248. The American *R* is Rochester, New York, 267,173, closely followed by Richmond, Virginia, 232,652.

S—Seoul, South Korea, 5,433,198. The American S may be a fooler. It isn't St. Louis, Missouri, San Antonio, Texas, or San Francisco, California. It is San Diego, California, 773,996.

T—Tokyo, of course, 8,442,634; with Tientsin, China, in second place 4,280,000 and Teheran, Iran in third place, 4,002,000. The American T is Toledo, Ohio, 367,650.

U—Ufa, Soviet Union, 923,000 (and that's not one that is likely to be guessed). The American U is Utica, New York, 91,340.

V—Vienna, Austria, 1,614,841. The American V is Virginia Beach, Virginia, 213,954.

W—Wuhan, China, 2,146,000, and Warsaw comes next 1,448,900. The American W is Washington, D.C., 711,518.

X—Xenia, Ohio, 25,373. Second is Xanthi, Greece, 25,341. This gives rise to a fine conundrum: "What have New York City and Xenia, Ohio, got in common?" The answer is that if you list the largest city in the world for each initial letter, the *only* American cities on the list are New York City and Xenia.

Y—Yokohama, Japan, 2,610,124. The American Y is Yonkers, New York, 192,509.

Z—Zaporozhye, Soviet Union, 760,000 (who'll get that?). The American Z is Zanesville, Ohio, 33,045.

Very well, then, if in another decade or so we're all still alive, I'll take another look at the situation.

17.
Nice Guys Finish First!

Naturally, I receive a good deal of mail from people I do not know who tell me what they think of me and my work. The vast majority of such mail is complimentary or, at the very least, polite. For this, I am endlessly grateful.

There is, however, a small group of letters that, for one reason or another, represent disapproval—and impolite disapproval at that. The problem then arises as to how to deal with these.

Alas, my reaction is invariably one of anger. Not, of course, at the disapproval (I don't expect to be approved of by one and all), but at the impoliteness, the sarcasm, the heavyhanded satire and so on.[1]

Well, then, I have worked out a system. In almost every case, I stop reading the letter when I realize it will anger

---

[1] Truth compels me to admit that I am sometimes impolite, sarcastic, and heavyhanded in my own writing, but I try increasingly not to be, and I think the tendency to be so is waning.

me, for I don't enjoy anger. Having stopped reading it, I naturally don't answer.

If, by any chance, I do finish the letter because it holds a horrid fascination for me, I strive, nevertheless, not to answer. I simply file it with my papers generally (which, for reasons known only to itself and the All-Knowing Computer in the Sky, Boston University collects with total lack of discrimination).

If I *must* answer, or blow a fuse, I write a sardonic and bitter answer, calling on my not inconsiderable supply of nonvulgar invective. Then I carefully place the answer in an envelope, seal it and stamp it. It is amazing how that discharges the venom. Of course, once the venom is discharged, there is no need to mail the letter. I tear it up, carefully destroying the stamp as I do so. (Unless I go to the full trouble of writing an answer, including the irrevocable demolition of a stamp, I do not feel the spiritual boil to have been properly lanced.)

If, by any chance, an answer should be absolutely required, the writing and destruction of the first letter makes it possible to write a second, milder letter or, even, if necessary, a third, still milder letter. When a proper level of politeness is reached, I mail it.[2]

I don't believe this behavior of mine is confined to me or is nothing but a peculiar and useless quirk. I think it's a rather general kind of reaction by the more civilized members of our species.

As a whole, we are all quick to anger and ache to meet each blow with a harder counterblow. As we grow older, wiser and more experienced, however, we reach the stage where we first judge whether the blow is really damaging, and if it is, we respond with the least energy required to disinfect the consequences.

This increase of mildness with age (or wisdom—the two

---

[2] Only once that I can remember was I so furious that I wrote a second letter still more insulting than the first, and then mailed it. It accomplished nothing, and there may be a moral to that.

are not identical, I suppose) is required of anyone who would aspire to the title of "a nice guy," and this is something to which I do indeed aspire.

And why do I aspire to it? Out of an inhuman virtue and saintliness?

Certainly not! I want the title out of selfish ambition. I happen to think that in the long run (despite Leo Durocher) nice guys finish first; and what I *really* want is to finish first.

Let me explain what I mean.

When we consider those animal species which have a level of intelligence high enough to make it appear to us that their actions are not motivated purely by instinct, we are tempted to personify. We wish upon them human methods of thought, imagining they freely choose or decide to do this or that.

Under these conditions, we sometimes cannot help feel chagrined, for it would appear that *Homo sapiens* is a particularly vicious species; that it alone among the animals fights needlessly to the death, that it alone seems to take pleasure in killing and in knowingly inflicting pain.

The "lower" animals, it would appear, fight over immediate matters of dispute; a specific piece of food, a specific mating opportunity, a specific territory. One of the competitors wins and one loses, the decision being reached with minimum violence, almost always far short of death or even of serious damage. Sometimes a conflict of threats is all that is needed. The loser leaves the scene and that conflict is ended.

Why are human beings so different? Why so combative? Why so deadly? And since they are, and since they rule the earth, is it a case of nasty guys finishing first?

Suppose we consider the difference between human beings and other animals. Surely there is a difference in intelligence to begin with. Human beings have considerably more brain power than other animals, and we may well be the only animals with a true time sense, as a consequence of that intelligence. Does that make a difference?

Well, when an animal fights for a specific object, he is a creature of the present only. If the specific object disappears, as when the competing animal manages to bolt down the food, or the prospective mate flees, or when the animal himself judges himself to be the loser and flees—it is all over. With the goal or the enemy or both out of sensory range, the reason for combat is over and neither the memory of past combat nor the anticipation of future combat serves to disturb the equanimity of the now-peaceful moment. (I don't say that a sufficiently intelligent animal doesn't remember or anticipate at all, but I think it does not do so sharply enough to disturb a peaceful present.)

Suppose, though, that intelligence advances to the point where the time sense becomes of major importance; when both memory and anticipation are strong and ever present.

In such a case, if X fights Y, X remembers past fights in which Y has troubled him or even perhaps thwarted him, and X anticipates further trouble of the sort in the future.

The intensity of the combat is bound to rise, then, as X strives not merely to drive Y away or to gain an immediate goal, but to inflict a defeat great enough to make up for past trouble and, perhaps, to ward off future trouble.

If, indeed, Y has given X *enough* trouble, then X may get ready to fight the moment it senses the presence of Y, even if there is no immediate goal that would make victory meaningful.

It may even be that the memory of past defeats may be so strongly and continually painful that X will, even without any cause whatever and without the presence of Y to act as a trigger, deliberately plan conflict in the future (under conditions favorable to himself) in order to restore the balance.

Even if X were victor in a combat, but only narrowly so, he might have the intelligence to anticipate the possibility of actual defeat the next time, and deliberately seek out conflict (under conditions favorable to himself) to inflict an overwhelming defeat once and for all—to inflict death, if possible, and end the problem.

In short, the growth of intelligence is bound to introduce new motives for conflict—shame, apprehension, desire for revenge or for security—all of which do not involve an immediate quarrel and cannot be satisfied by a minimal victory.

It may be, then, that human beings are nastier than other animals, not because they are reasonlessly nasty, but because they are more intelligent than other animals. Intelligence itself inevitably heightens the viciousness of conflict.[3]

There is another quality that seems to go along with increasing intelligence and that is an increasing ability to bend the universe to one's will by taking advantage (wittingly or unwittingly) of the way in which the universe works. Intelligence, to put it another way, can imply the development of a technology.

It takes considerable intelligence to do this on a significant scale and only *Homo sapiens*, in the more than 3-billion-year history of life on earth has developed enough intelligence (in either quantity or quality or both) to develop a significant technology.

By technology, human beings develop tools to extend and refine their natural abilities to influence their environment and put these tools at the service of their propensity for violence against each other.

Conflict between human beings becomes not merely the interplay of arms, feet, heads, teeth and nails, but of rocks, bones, clubs, knives, spears, arrows and so on indefinitely.

It is plain to see that the more intelligent a species becomes and the longer it remains intelligent, the greater the damage it can do to its own members, to other species, and to the world generally.

Naturally, human beings, as they grow more intelligent or

[3] What about the dolphins, which are intelligent and yet peaceful? We don't know them well enough, yet, to be sure of the true extent of either characteristic.

gather greater experience or both, can learn to repair the damage done by conflict and to do so more rapidly and effectively.

Yet does the ability to repair keep step with the ability to destroy? It would seem that as technology grows more complex, it also grows more vulnerable, so that even as the ability to destroy increases rapidly, the difficulties of repair also mount. From that, we might deduce that the capacity for destruction is bound to outstrip the capacity for repair. Sooner or later, then, the technology will be destroyed, probably civilization along with it and, just possibly, humanity itself as well.

Even if humanity survives, it may lack the capacity to restore an advanced technology because of the disappearance of cheap and easy sources of energy. And if we do restore an advanced technology, it will only be to court destruction again.

The final conclusion would seem to be, then, that the kind of intelligence that leads to technology is self-limiting and even self-destructive not only on earth but, presumably, on any world in which such an intelligence has evolved.

The universe, in that case, may have witnessed the rise of countless civilizations which are now all dead, except for a few like ours that are too young yet to have died but are fated to do so soon. (I referred to this briefly, by the way, in "Where Is Everybody?", *F & SF*, December 1978.)

And perhaps uncounted numbers of civilizations are yet to arise during the remaining lifetime of the universe, only to die quickly in their turn.

Nasty guys, in other words, whatever their short-term success, in the long run finish last.

But wait . . .

So far I have only considered the nastiness of *Homo sapiens*, its propensity for cut-throat competition and conflict. Are there no compensating factors that can alter the dismal conclusion I have just reached?

For one thing, not all competition and conflict is bad.

There is the useful competition to reach a creditable goal first for some reward that does not imply direct physical harm to the loser.

Even malignant conflict has its beneficial side effects. In the crucible of war, great self-sacrifice is demanded and obtained while the records seem to show that the arts and sciences flourish in times of stress.

Yet surely that is not enough. If conflicts grow steadily more deadly with time, the point is bound to be reached where no possible beneficial side effect of the spirit of competition can possibly make up for the destruction.

Still, in talking about the self-sacrifice demanded in war, I'm talking about co-operation. If conflict is "nasty" then co-operation is "nice," and surely *Homo sapiens* has the capacity for "niceness," too.

Just as not all competition is malignant, so not all co-operation is beneficent. The co-operation of the beehive or the anthill, which destroys the capacity for individual initiative and creativity and limits the potentiality for diversity, may keep a species alive but slows, or even stops, the growth of a technology. If such a co-operation does not lead to death, it may lead to death in life, which is not much better.

The type of co-operation fostered by conflict, which must be aimed entirely at making victory more probable, is rather beehivish in nature, as anyone who has been in the armed forces can attest, and is not my idea of beneficent co-operation.

Is it possible to have co-operation, but of a looser kind, that leaves room for individuality and even for nonmalignant competition?

Perhaps co-operation does not arise as easily as competition does, but competition leads to war and I have already said that war leads to co-operation of a sort. Even primitive war does.

If a species is intelligent enough to have memory and foresight, individuals who have suffered defeat or who fear future defeat may see the value of seeking allies. Thus, if X is defeated by Y, X and Z together may nevertheless defeat Y.

The development of the notion of co-operation is not just a likelihood but a virtual certainty, at least for *Homo sapiens*. While gorillas and orangutans are loners, chimpanzees are tribal, and undoubtedly, the hominids were tribal from the start.

Tribes have other uses than self-defense even among animals of only moderate intelligence. They can become hunting bands, for instance.

One human being, even armed with a spear or a bow and arrow, can do nothing to a mammoth but watch from a very safe distance. A co-operating group of human beings, armed each with similar primitive weapons, can destroy a mammoth, and, indeed, such hunting groups managed, long before the birth of civilization, to drive these magnificent creatures to extinction—as well as other large, but insufficiently intelligent species.

Of all tribal species, only *Homo sapiens* developed a technology, and, as it happens, there is very little in the way of technology that a single human being, starting from scratch, can develop. A group of human beings, with diverse talents, are much more likely to have the succession of ingenious ideas that bring about the growth of technology.

Not only that, but the growth of technology seems to require, inevitably, the development of larger and larger co-operating groups to maintain that technology at its existing level and to bring about further growth.

The development of agriculture required a large population of farmers not only to till the fields and weed and hoe and sow and reap and do all the work required to produce a year's supply of food, but also to make the implements needed, to construct and maintain the irrigation ditches, to build walled cities and collect armaments to protect themselves from surrounding tribes who, not having sown, would be glad to collect the reapings by force.

Fortunately, the development of agriculture made it possible to support a larger population than would have been possible without it. In general, it has been true that advances in technology have both produced and used a larger and denser population than before.

To make the technology work, moreover—and this is the crucial point—there must be co-operation at least over a political unit large enough to be economically useful. Through history, as technology has advanced, the size of these economic units has necessarily increased from tribal patches, to city-states, to nations, to empires.

Within these units co-operation has been brought about, despite the natural tendency to destructive competition, by the application of governmental authority, internal police and, most of all, the strictures of custom, social pressure and religion.

The general advance in the size of the units within which co-operation is maintained has, at the present day, produced governmental control over a population of 950 million people in China; 22 million square kilometers of area in the Soviet Union; and one third of the real wealth of the world in the United States.

The advance has not been smooth and steady. The stresses of internal decay and external pressure have led to the fall of empires and the periodic destruction of central authority and its replacement by smaller units. Such periods of regression usually result in a "dark age."[4]

Today, the world undergoes centrifugal decomposition

---

[4] There are people who, disturbed by "big government" today and its tendency to curb the advantages they might gain if their competitiveness were allowed free flow, demand "less government." Alas, there is no such thing as less government, merely changes in government. If the libertarians had their way, the distant bureaucracy would vanish and the local bully would be in charge. Personally, I prefer the distant bureaucracy, which may not find me, over the local bully, who certainly will. And all historical precedent shows a change to localism to be for the worse.

politically, as the old European empires break up and as cultural minorities demand nations all their own; but economic units continue to grow larger and the only economic unit that makes sense today is the whole planet.

In one way, it's the political units that count, for it is they who wage war. Though peace is maintained within the units (if we ignore endemic crime and violence, and occasional terrorism, rebellion and civil war) there is war between them.

City-states warred against each other interminably in ancient Greece and in Renaissance Italy; feudal estates did so in medieval Europe and early modern Japan; nations did so in early medieval China and modern Europe, and in all cases until modern times there were conflicts with barbarians on the fringes.

The intensity and destructiveness of the conflicts shows a general rise with advancing technology, so that despite the growing size of the units within which co-operation can be counted on, competitiveness may still win out. Destruction still threatens to outpace the capacity for recovery.

We now live at a time when the outcome clearly hangs in the balance. One more all-out general war and civilization will probably be destroyed—possibly for good.

Indeed, even if the realization of this keeps the war from happening, the existence of potential conflict keeps the minds and energy of all the competing nations on each other as the enemy and *not* on those true enemies which threaten us all—overpopulation, resource depletion and technological inadequacy.

Nasty guys will finish last.

How to prevent that? We have reached the point where we can no longer afford armed competition; nor can we afford to have competition preoccupy us so that we cannot truly co-operate to solve global problems. There must be sufficient international co-operation to serve as the equivalent of a world government (though that should entail as much local autonomy as is consistent with global success).

This is needed not only to avoid destruction, but to allow technology to continue to grow and improve. The time has come when projects are possible which can use and, indeed, must use the whole effort of the global economy and population. To solve our problems involving population, energy, pollution (yes, even a peaceful technology has destructive side effects that must be reversed) a global effort is required and I believe that in every case the penetration and exploitation of space as the enlarged sphere of human activity is essential.

It is my feeling that civilization will not survive and space will not be conquered without a working global co-operation among the nations, and that it is possible for the peoples of the earth to choose to indulge in such a co-operation. They may *not* choose to, but they can if they wish to. If they want to be nice guys, they can be, and nice guys will finish first.

I can also maintain the converse. I believe that any planetary civilization which reaches the stage of space exploration and exploitation will have learned to handle the tendency to destructive conflict that, so far, has seemed inseparable from intelligence. They will have learned to be nice guys.

If they haven't, they will have remained bound to their planetary surfaces and will have decayed. In fact, in all likelihood, they will have destroyed themselves.

It is for this reason that I do not fear contact with extraterrestrial civilizations. If we get to them, we will be stronger, and have nothing to fear (nor they from us since we will be peaceful people). If they get to us, and *they* are strong, *they* will be peaceful people.

Yet can we be sure? Might not a civilization that has learned to live at peace with itself yet not hesitate at conflict with an extraplanetary civilization? Might they not even welcome a chance to exercise their repressed delight in destruction?

Why should they? That is judging a true civilization from the standpoint of our own barbarism.

For instance: We have one case of a superior civilization crossing space to visit other, possibly life-bearing worlds. The case of ourselves.

Our instruments have landed on the moon, on Mars, on Venus, and have made close approaches to Mercury, Jupiter and Saturn. In the case of the moon, at least, some of the instruments contained human beings.

What's more, the intruders on or near other worlds are not individuals from a planetary civilization, but are ourselves, *Homo sapiens*, from a world full of conflict, hatred and destructiveness.

How have we behaved? For one thing, the nations of the earth, notably the United States and the Soviet Union, have behaved with surprising co-operation in the space effort. Each maintains spy satellites and there is talk (only talk) of killer satellites, but there is co-operation, free flow of information and no sign of attempt to do deliberate harm.

And how did we behave toward possible life on other worlds? We acted with the greatest circumspection. We sterilized our vessels at enormous expense so that we might not unwittingly introduce earthly organisms that might harm any native life. We *protected* that life with all our might even though we knew it was most sure to be unintelligent and primitive, and very likely was not there at all.

Yes, we did it out of self-interest. We wanted to study those life forms to gain knowledge and perhaps to turn that knowledge to our own benefit.

That, however, is equivalent to saying that altruism is to our own long-term selfish benefit—which is exactly what I've been saying all through this essay.

Nice guys finish first.

Biological evolution teaches it and human history teaches it and whether we learn it *in time* is the great question of the moment.